Contents

Introduction to APP for Maths   2

Year 2 and Year 1 task-mapping chart   4

Assessment focus coverage chart   5

Task 1   Counting on and back in 1s and 10s   6

Task 2   Estimating   8

Task 3   Tens and units   10

Task 4   Compare numbers   12

Task 5   Partitioning   14

Task 6   Fractions   16

Task 7   Understanding doubling and halving   18

Task 8   Missing numbers   20

Task 9   Adding three numbers   22

Task 10 Mental strategies   24

Task 11 Subtracting 10s and near 10s   26

Task 12 Multiples of 2, 5 and 10   28

Task 13 Strategies for doubling and halving   30

Task 14 Addition and subtraction   32

Task 15 Subtracting   34

Task 16 Addition and subtraction: problems with money   36

Task 17 Division as the inverse of multiplication   38

Task 18 Addition and subtraction problems   40

Task 19 Recording number problems   42

Task 20 Shape properties   44

Task 21 3D shapes   46

Task 22 Position and movement: positions   48

Task 23 Position and movement: routes   50

Task 24 Measuring and comparing length using centimetres   52

Task 25 Recognising right angles   54

Task 26 Mass   56

Task 27 Reading scales   58

Task 28 Collecting data   60

Task 29 Sorting and classifying   62

Task 30 Interpreting a block graph   64

Pupil Sheets   66

Introduction to APP for Maths

APP for Maths Year 2 contains 30 tasks. It covers the assessment foci for Levels 2 and 3, with the main emphasis on L2.

The tasks in this book are designed to support the periodic assessment that is a key part of the 'Assessing Pupils' Progress' (APP) evidence gathering. The tasks help you, the teacher, to gather evidence to support your own professional understanding of each child's level of attainment.

Gathering assessment evidence to address assessment foci

The tasks have been designed to give you a good understanding of the National Curriculum level at which children are working. They have been written to work in conjunction with the Assessing Pupils' Progress materials, alongside the assessment guidelines and assessment foci (AFs). They should be used when children have already been working on the same or similar topics. They have been designed for children working in a small group with a teacher. It is envisaged that each child will work in a group on several tasks over a term. Not all children need to do every task: the tasks should be regarded as a bank of activities to be used selectively.

How effective these assessment tasks are at determining an accurate level for each child depends very much on the way that they are used. Gaining a secure understanding of the level at which each child is operating can only be achieved by using them in the ways outlined, so it is important to go through them carefully beforehand, and to read the information below.

The tasks should be used alongside the QDCA and National Strategy APP Guidance and Standards files.

Using the tasks in the classroom

Each assessment task is linked to two assessment foci:

- one within the content attainment targets: Ma2 Number, Ma3 Shape, Space and Measures, Ma4 Handling Data; *and*

- one within the three assessment foci for Ma1 Using and Applying Mathematics (Problem Solving, Communicating and Reasoning).

This will enable you to assess Ma1 in an integrated way alongside the core content aspects of mathematics.

The **Teacher Sheet** for each task shows at a glance what is being assessed, and anything you will need. The tasks are linked to the learning objectives from the PNS *Framework for mathematics* so that you can see quickly when it might be appropriate to use them. A task overview lets you see quickly what the task involves. Activity notes explain how to introduce the task, what children should do and what to discuss with the group. Every task has an Extension to help you gather further evidence at Level 3. Questions to ask, and help with what to look out for as children are working, are covered by a final 'Observe and ask' section.

The **Assessment Sheet** for each task contains a grid showing how children may demonstrate understanding within the Using and Applying Mathematics assessment focus and the mathematical content assessment focus at Levels 2 and 3. Specific examples of what a child might say or do will help you make judgements about levels more easily.

These grids can be photocopied and highlighted and, where appropriate, examples of the child's work attached as evidence. This can be kept as part of an assessment portfolio. When preparing to use the tasks, read the assessment criteria pages so that you have an idea of what sorts of responses to look out for.

Most, but not all tasks have an accompanying **Pupil Sheet**.

In order to support differentiation, the table on page 4 maps the tasks in this book to those in *APP for Maths Year 1*. Where the demands of these tasks are too high, you can find an equivalent task from the Year 1 book. The tasks do not always cover the same content, but they do address the corresponding assessment foci, with a stronger emphasis on Level 1. You will find *APP for Maths Year 3* useful for more able children who need further assessment of the Y3 objectives at Level 3.

The tasks have the flexibility to be used in several different ways, such as:

• for small group, adult-led activities during the main part of a lesson;

• using one task with different groups on different days during a block of work;

• using specific tasks during lessons in which assessments take place;

• with targeted children only. This is especially useful with children who are performing better than or not as well as expected in more formal tests. In this context, the assessment tasks provide complementary evidence to help the moderation of the test result.

Making assessment judgements

No single task can determine that a child is low, secure or high Level 2. However, observing a child working on several tasks over a period of time will provide evidence of how they are functioning at a particular level mathematically and should also give an indication of their security within that level. If a child is successful in a number of tasks at a particular target level, you will need to decide how consistently the evidence fits the criteria in the assessment focus in order to determine whether the child's performance at that level is 'low', 'secure' or 'high'.

Children can work in a group but it is essential to note individual responses. Encourage them to explain what they are doing or to talk through an answer, and then observe closely how they work on the tasks. These are all crucial elements in determining how well the evidence supports assessment. It is through appropriate questioning and prompts when children are giving explanations that tasks provide evidence for Ma1 Using and Applying Mathematics alongside other assessment foci.

A child may prove to be at one level for Ma2, Ma3 or Ma4 and a lower one for Ma1 Using and Applying Mathematics. This may happen where they are able to show factual knowledge of some mathematics but are less able or confident in discussing or explaining it. Or in the case of problem solving, they might need a degree of support in choosing how to go about a problem. It is unlikely that children will be working at a higher level for Using and Applying Mathematics.

Sometimes tasks will give evidence of other assessment foci than those targeted; make a note when this happens and use this as evidence for those assessment foci too.

Year 2 and Year 1 task-mapping chart

Use this chart to help you choose assessment tasks from *APP for Maths Year 1* for children working at a lower level.

	Y2 TASKS (L2/3)		Y1 TASKS (L1/2)
1	Counting on and back in 1s and 10s	2 4 7	Reading numbers and counting on 1 more Counting on and back in tens Number sequences
2	Estimating	1	Counting
3	Tens and units	5	Tens and units
4	Compare numbers	3	Comparing and ordering
5	Partitioning		
6	Fractions	8	Fractions
7	Understanding doubling and halving	13	Doubles
8	Missing numbers	9	Counting back
9	Adding three numbers	12	Counting on to bridge ten
10	Mental strategies	11	Addition and subtraction
11	Subtracting 10s and near 10s	15	Subtract by counting back to a multiple of ten
12	Multiples of 2, 5 and 10	6	Counting in twos, fives and tens
13	Strategies for doubling and halving	13	Doubles
14	Addition and subtraction	16	Addition number sentences
15	Subtracting	10	What is the difference?
16	Addition and subtraction: problems with money	17 18	Exchanging coins Solving problems with money
17	Division as the inverse of multiplication		
18	Addition and subtraction problems	19	Adding 2 or 3 numbers
19	Recording number problems	20	Checking number sentences
20	Shape properties	21 23	Problem solving: shape sequences Properties of shapes
21	3D shapes	22	3D shapes
22	Position and movement: positions	24	Position and movement
23	Position and movement: routes		
24	Measuring and comparing length using centimetres	25	Length
25	Recognising right angles		
26	Mass	27	Mass
27	Reading scales		
28	Collecting data	28	Sorting shapes
29	Sorting and classifying	29	Interpreting data
30	Interpreting a block graph	30	Creating a pictogram

Assessment focus coverage chart

AF/Level	AF1 Using and applying mathematics			AF2 Number						AF3 Shape, space and measures			AF4 Handling data and Using and applying mathematics	
	(i) Problem solving	(ii) Communicating	(iii) Reasoning	(i) Numbers and the number system	(ii) Fractions and decimals	(iii) Operations, relationships between them	(iv) Mental methods	(v) Solving numerical problems	(vi) Written methods	(i) Properties of shape	(ii) Properties of position and movement	(iii) Measures	(i) Processing and representing data	(ii) Interpreting data
L3	2, 7, 13, 17, 19, 24, 25, 26, 27, 28	1, 5, 6, 9, 14, 15, 16, 18, 20, 21, 22, 23, 30	3, 4, 8, 10, 11, 12, 29	1, 2, 3, 4, 5	6	7, 8	9, 10, 11, 12, 15	13, 14, 16, 17	18, 19	20, 21	22, 23	24, 25, 26, 27	28, 29	30
L2	2, 7, 13, 17, 19, 24, 25, 26, 27, 28	1, 5, 6, 9, 14, 15, 16, 18, 20, 21, 22, 23, 30	3, 4, 8, 10, 11, 12, 29	1, 2, 3, 4, 5	6	7, 8	9, 10, 11, 13, 15	12, 14, 16, 17	18, 19	20, 21	22, 23	24, 25, 26, 27	28, 29	30

Task 1 Counting on and back in 1s and 10s

Framework objectives • Count on or back in 1s or 10s from 2-digit numbers.	**Assessment foci** AF1 – Using and applying (communicating) AF2 – Number (numbers and the number system)
Task overview Children find numbers 1, 10 or 20 more or less than a range of 2- and 3-digit numbers, with or without the support of a 0–99 number square.	**Resources** • Pupil Sheet 1.1 for extension activity, one per child • 0–99 number square • Individual whiteboards and pens

Activity

• Revise counting to and back from 90 in 10s on the number square. Then draw a square in the middle of the board and ask children to copy it onto their whiteboards. Write a 2-digit number in the middle of the square (for example, 37) and tell children to copy this as well. Then ask children to draw a second square directly below the first and to write the number that is 10 more in the square (47).

• Repeat with a number 10 less/one more/one less than 37. When finding 10 less, ask: *What number have you have written in your square? How many 10s is that? How many 1s?* Write the number 27 (i.e. 10 less than 37) and ask: *How many 10s will it have? Which digit stays the same? Why does it stay the same?* Children should check their answers using the number square.

• Repeat the activity, this time without the support of a number square.

Extension

• Children should work through Pupil Sheet 1.1 to find numbers 10 and 20 more/less than a range of 2- and 3-digit numbers, including crossing 10s and 100s boundaries.

Answers
1a 103; **b** 157; **c** 206; **d** 299
2a 108; **b** 154; **c** 267; **d** 412
3a 68; **b** 96; **c** 267; **d** 293
4a 94; **b** 248; **c** 285; **d** 399

Observe and ask

• *When we count up in tens, what happens to the number?*
• *Which digit changes – the tens or the units?*
• *Why doesn't the units digit change?*
• When finding 10 more than 37, do children write *47* without hesitation or do they write *38*?

Task 1 Counting on and back in 1s and 10s

	AF1 – Using and applying Communicating	AF2 – Number Numbers and the number system
Children working at L2	• **Discuss their work using mathematical language, for example, with support.** Children should be able to explain how the number changes as 1 or 10 is added or subtracted with reference to moves on the number square.	• **Recognise sequences of numbers.** Children should be able to say the number 1 or 10 more than any 2-digit number up to 90. Children may find it more difficult to say the number 1 or 10 less, especially when bridging over a decade. Some may continue to need the support of a number square.
Children working at L3	• **Discuss their mathematical work and begin to explain their thinking.** Children should be able to discuss the activity using appropriate place-value vocabulary to explain which digit changes and why.	• **Recognise a wider range of sequences.** Children will be able to find numbers that are 20 more or less than the original number, and both 10 or 20 more or less than any number over a wider range of numbers, including bridging over decades and hundreds.

Task 2 Estimating

Framework objectives
• Estimate a number of objects up to 50.

Assessment foci
AF1 – Using and applying (problem solving)
AF2 – Number (numbers and the number system)

Task overview
Children estimate the number of objects in a container and devise strategies for counting the total accurately and efficiently.

Resources
• Counters, cubes, conkers or pasta pieces
• Plastic (see-through) jugs or other containers
• Sugar paper and pens

Activity

• Prepare several see-through containers containing between 30 and 60 objects (use either counters, cubes, conkers or pasta pieces but not a mixture of these). Ask each child to estimate the number of objects in one of the containers. Write down each child's estimate.

• Carefully pour and spread out the objects onto a piece of sugar paper. Ask children if they want to change their estimate and if they can suggest a way to count the objects. (Note that since problem solving is one of the foci of this assessment activity, leave it to the children to decide what would be the best way to do this.)

• After the counting is complete, give each child one of the other containers and ask them to repeat the activity – i.e. jot down an estimate, pour the objects onto sugar paper and then find an efficient way to count them. Ask children to arrange the objects on the sugar paper in such a way that it makes it easy to count.

Extension

• Give children a larger container and ask them to estimate to the nearest ten how many cubes or conkers would fit in it. Alternatively, use the same containers but with smaller cubes or other objects, such as paper clips or buttons.

Observe and ask

• *How many objects do you think there are?*
• *What would be a good way to count these?*
• *How could you check easily?*
• What strategies do children have for counting?
• How accurate are children's estimates?

Task 2 Estimating

	AF1 – Using and applying Problem solving	AF2 – Number Numbers and the number system
Children working at L2	• **Select the mathematics they use in some classroom activities, for example, with support.** Children may need prompting to use the types of strategies described opposite but will appreciate the value of using them and show some evidence of using them independently with further examples.	• **Count sets of objects reliably.** Children should be able to make a reasonable estimate of the number of objects in a container – within 10 of the actual number is reasonable. They should have some strategies for counting efficiently (for example, counting in 2s or 5s). Children should also have strategies for arranging the objects on the sugar paper (for example, in groups of 5 or 10 or in straight lines), which make it easier for them to check their estimate.
Children working at L3	• **Select the mathematics they use in a wider range of classroom activities, for example, with support.** Children will make connections to previous estimating and counting activities. They will draw upon place-value understanding to appreciate that, when estimating larger numbers of objects, getting the exact number right is unlikely so an approximation to the nearest ten is an acceptable level of accuracy.	• **Use place value to make approximations.** Children should realise the need to have systematic approaches to counting the objects, which they will be able to articulate. Their estimates will be thoughtful (for example, they may think about how many cubes fit across the container and the number of layers of cubes there are).

Task 3 Tens and units

Framework objectives
• Know what each digit in a 2-digit number represents, including 0 as a place holder.
• Partition 2-digit numbers into T and U.

Assessment foci
AF1 – Using and applying (reasoning)
AF2 – Number (numbers and the number system)

Task overview
Children demonstrate their understanding of place value by matching 2- and 3-digit numbers to representations of these, using different resources.

Resources
• 30–99 number cards
• £1, 10p and 1p coins
• 0–99 number square
• Hundreds, tens and units place-value cards, one set per child

Activity
• Shuffle the number cards 30–99 and give one to each child. Ask children to read out their number and say how many tens and how many units it has. Challenge them to make the matching amount with 10p and 1p coins. Check that they do match. (Note that the coins provide a context for manipulating number – there is no need to focus on the monetary amount.) If necessary, refer to a number square to help by saying: *Point to your number. Now point to the number 10 bigger.*

• Repeat, making the numbers 10 bigger/smaller, then 1 bigger/smaller. Ask children to say which coin they would use to make their number 10 bigger.

• Then give each child a set of tens and units place-value cards and ask them to make 20 from their cards. Ask:
  ○ *How many 10s do you need?*
  ○ *How many 1s do you need?*
  ○ *How can you make your number 10 bigger?*
  ○ *What is your new number?*
  ○ *How will you make your number 6 bigger?*

Extension
• Use place-value cards to make 3-digit numbers, and ask children to match £1, 10p and 1p coins to the cards. Repeat, increasing/decreasing numbers by 100, 10 and 1.

Observe and ask
• *What is your number?*
• *Which number is the tens number? Which is the units number?*
• *How will you match the 10p and 1p coins to the number on your card?*
• Do children count the 10p coins first?
• Do children write the number that is 10 bigger correctly?
• Can children solve problems that involve bridging the tens?
• *Can you make the number 1 bigger? Which digit will you change?*
• Do children add an extra unit when making a number 1 bigger?
• *Take 20 away. What is your new number?*
• *Is your new number bigger or smaller than your original one?*

Task 3 Tens and units

	AF1 – Using and applying Reasoning	AF2 – Number Numbers and the number system
Children working at L2	• **Explain why an answer is correct, for example, with support.** Children should be able to locate numbers on a number square and efficiently match 2-digit numbers to different representations, for example, coins or place-value cards.	• **Begin to understand the place value of each digit.** Children will be able to read the number cards and should be able to tell you how many tens and units are in each number, pointing to the correct digit when saying tens and units digits. Children should know that in multiples of ten, the zero establishes that there are only tens and no units. They may sometimes get confused when the digits of the original number are the same (for example, 44).
Children working at L3	• **Review their work and reasoning.** Children should be able to explain why different digits change when either 1, 10 or 100 is added or taken away, confidently using place-value vocabulary.	• **Understand place value in numbers to 1000.** Children will be able to confidently complete this task for 3-digit numbers, identifying what value each digit represents and matching these to their representations with different resources.

Task 4 Compare numbers

Framework objectives

• Compare two or more 2-digit numbers; introduce the greater than (>) and less than (<) signs.
• Say a number lying between two numbers, up to at least 100.

Task overview

Children read, write, order and compare a range of numbers.

Assessment foci

AF1 – Using and applying (reasoning)
AF2 – Number (numbers and the number system)

Resources

• Pupil Sheet 4.1 for extension activity, one per child
• 1–100 number cards
• Post-it notes
• 0–100 number line

Activity

• Shuffle the number cards 1–100, and give each child four cards. Ask them to arrange their numbers in order. Children should take turns to read their order aloud. Invite each child to choose the largest and smallest number and put aside the other two cards. Then ask them to write an 'in-between' number on a Post-it note and to place it between their two cards. Shuffle the cards and repeat, giving each child six cards this time. A 0–100 number line could be made available for children to refer to.

Extension

• Write on the board a row of six numbers between 50 and 550, for example, the row given in question 1 on Pupil Sheet 4.1. Check that children are able to read each number. Ask:
  ○ *How will you order your numbers?*
  ○ *How do you know that you have ordered them all?*
  ○ (Point to a number in a child's order.) *How do you know that this number is in the right place?*
  ○ *Write a number that comes between these two numbers.*

• Children should then complete Pupil Sheet 4.1 independently.

Observe and ask

• *How will you arrange your cards in order?*
• Do they arrange the cards correctly?
• *Did you find any cards difficult to order? Which cards? Why?*
• *Can you explain why you put this card here?*
• *Which card was the most difficult to position? Why?*
• *Can you write a number that comes between your two cards? Is there another number you could have written?*
• Do children look at the tens numbers first?
• When comparing numbers in the same decade, do children compare the units digits?

Task 4 Compare numbers

	AF1 – Using and applying Reasoning	AF2 – Number Numbers and the number system
Children working at L2	• **Explain why an answer is correct, for example, with support.** Children should be able to identify other numbers between two numbers and realise, some with prompting, that (in most cases) more than one answer is possible to this question. They should draw upon place-value understanding to confirm their solutions.	• **Begin to understand the place value of each digit; use this to order numbers up to 100.** Children will be able to arrange their cards in order and be able to explain their reasons for arranging them, for example, by referring first to the tens and then the units. Some children may confuse 'in-between number' with 'halfway between'.
Children working at L3	• **Review their work and reasoning.** Children are likely to demonstrate strategic ways of ensuring they have ordered the numbers correctly; for example, looking across the numbers, identifying the numbers in turn and crossing them out as they record them in order, then reading them through, checking for any errors.	• **Understand place value in numbers up to 1000.** Children will confidently and independently tackle the extension task, demonstrating place-value understanding of 3-digit numbers. Children's good understanding of place value should ensure they do not get confused by similar-looking/sounding digits in the number sets.

Task 5 Partitioning

Framework objectives
• Read and write numbers up to 100 in figures.
• Begin to partition 3-digit numbers into H, T and U.

Task overview
Children demonstrate their understanding of partitioning 2- and 3-digit numbers using a range of resources.

Assessment foci
AF1 – Using and applying (communicating)
AF2 – Number (numbers and the number system)

Resources
• 0–99 number square
• Post-it notes
• Hundreds, tens and units place-value cards
• Multibase apparatus (flats, longs and units)

Activity

• Pin up the number square and point to a number, for example, 47. Ask how many tens there are, and how many units. Repeat for 99 – ask children to write down which number comes next. Check children's answers.

• Write 100 on a Post-it note and place it on 0 on the number square. Ask children to add a hundred to numbers on the number square. Repeat up to an appropriate number, such as 124.

• Give children the hundreds, tens and units place-value cards and ask them to make 3-digit numbers.

Extension

• Ask children to use both place-value cards and multibase apparatus to make 3-digit numbers.

• Next ask children to make these numbers larger or smaller by 2- or 3-digit numbers of increasing difficulty, by changing the appropriate digits.

Observe and ask

• *Look at 25. How many tens does it have? How many units?*
• *Who can write 125?*
• *How is it different from 25?*
• *Is there an extra digit? What do you think this new digit could be? (A hundred)*
• *How much bigger than 47 is 147? How do you know?*
• *Can you show me how to make (for example) 367 using your place-value cards?*
• *How did you make this number? In what order did you gather your cards?*
• *Now make the number 1/10 bigger. Which digit changes?*

Task 5 Partitioning

	AF1 – Using and applying Communicating	AF2 – Number Numbers and the number system
Children working at L2	• **Begin to represent their work using symbols and simple diagrams, for example, with support.** Children should, some with prompting, be able to relate 2- and some 3-digit numbers to different physical representations such as multibase equipment. They will be able to use place-value cards to make the numbers and demonstrate the value of each digit.	• **Begin to understand the place value of each digit.** Children will be able to partition 2-digit numbers and know what each of the digits represents. They will be able to add 100 to the numbers using the activity prompts, although actually saying and writing the numbers from 101 to 120 may be tricky for some. Some children may be beginning to read and write 3-digit numbers but not always be sure what each digit represents.
Children working at L3	• **Use and interpret mathematical symbols and diagrams.** Children will not need a physical resource to support with this activity but, when asked, will clearly be able to show how the different physical resources relate to understanding the value of each digit.	• **Understand place value in numbers to 1000.** Children have a good grasp of place value and can identify value of digits in 3-digit numbers – some will be able to go beyond this to 4-digit numbers. Children will not have any problems reading or writing numbers where the zero is used as a place holder, for example, 207 or 340.

Task 6 Fractions

Framework objectives
- Begin to recognise halves and quarters of small numbers of objects.
- Recognise fraction notation.

Task overview
Children demonstrate their understanding of fractions as parts of a whole, and fractions of numbers in a practical context.

Assessment foci
AF1 – Using and applying (communicating)
AF2 – Number (fractions and decimals)

Resources
- Paper circles (to represent cakes)
- Red counters (to represent cherries)

Activity

- Ask each child to fold a paper circle in half, checking that the edges fit exactly. Explain that the whole circle (or cake) has been divided in half. Write *half* on the board and ask the children to write *half* on one side of the folded paper circle. Ask how many halves there are in the whole circle.

- Write $\frac{1}{2}$ on the board and explain the notation. Hold up the semi-circle to indicate half the cake, then open the circle to show the two halves. Tell children: *There are two halves in a whole cake. Take 6 red counters (cherries) and share them so that both halves of your cake have the same number of cherries.*

- Repeat for other even numbers up to 20 cherries. Then repeat the activity using 12 cherries only, but this time folding the 'cake' in half twice so that it is divided into quarters.

Extension

- Ask children to halve bigger numbers, for example, 42, 48. Ask them to explain their working.

- You could also extend the work on quarters up to 24 cherries.

Observe and ask

- Do children all fold the paper exactly in half?
- Do children all understand the fraction notation $\frac{1}{2}$?
- *How many halves are there in the whole cake? Can you show me half of the cake?*
- *Share 12 cherries so that each cake half has the same number. How many cherries on each half of the cake? How many on the whole cake?*
- Can children write: *half of 12 is …*?
- *How many quarters make half? Are two quarters the same as one half?*

Task 6 Fractions

	AF1 – Using and applying Communicating	AF2 – Number Fractions and decimals
Children working at L2	• **Discuss their work using mathematical language, for example, with support.** Children should be able to talk through the activity using the appropriate vocabulary of fractions in both the context of halving of a shape and of halving a number of objects.	• **Begin to use halves and quarters; relate the concept of half of a small quantity to the concept of half of a shape.** Children will be able to halve the paper, although some may need help with putting the edges exactly together. They should understand the relationship between two halves and a whole. Children should be able to recognise and write simple fractions (for example, $\frac{1}{2}$, $\frac{1}{4}$) using appropriate notation. They should be able to answer questions about how many cherries are on each half and the whole cake and be able to share the cherries between the four quarters when the number given is a multiple of 4 (for example, 12, 16 or 20).
Children working at L3	• **Discuss their mathematical work and begin to explain their thinking.** Children will be able to articulate their knowledge of fractions confidently; for example, knowing that the denominator ('bottom') of the fraction tells how many parts to divide a shape or a number into. Implicit in this is an understanding of the relationship between fractions and division.	• **Use simple fractions that are several parts of a whole and recognise when two simple fractions are equivalent.** Children will have a range of mental strategies that they are able to call on (for example, partitioning, multiplication facts, use of place-value understanding) to support finding a half and a quarter of different numbers. Children will have a good grasp of the different models of fractions and how they connect.

Task 7 Understanding doubling and halving

Framework objectives
• Recognise halving as the inverse of doubling.

Task overview
Children will have opportunities to double and halve a range of numbers with or without the support of different resources, showing understanding of the relationship between the operations.

Assessment foci
AF1 – Using and applying (problem solving)

AF2 – Number (operations, relationships between them)

Resources
• 10p, 5p and 1p coins
• 1–50 number cards

Activity

• Use your fingers to double/halve up to ten. Put your thumbs together and say: *One plus one is two.* Then pull them apart, saying: *Half of two is one.* Repeat up to five plus five.

• Then give children two 5p and eight 1p coins. Ask them to make amounts up to 18p – for example, 6p + 6p (5p + 1p and 5p + 1p) and 9p + 9p (5p + four 1p coins and 5p + four 1p coins). Always say the inverse; for example, *If 7p + 7p (double 7p) is 14p, what is half of 14p?*

Extension

• Children should take turns to pick a number card from 1–50. If they can say the doubling and halving sentences correctly, they win 20p. For example, if a child picks 24, he/she should say: *Half 24 is 12; double 12 is 24.* (Note that if children pick an odd number card, they should always double the number first, then halve it.)

• Next write the numbers *22, 24, 26* on the board. Ask children to double each number and to describe their strategies. Ensure coins are available to help them. Do they automatically use existing knowledge to work out the answer? Some will use coins because they are there; others will double, having partitioned first.

Observe and ask

• *Can you show me double four (four plus four) with your fingers? Can you write the answer?*
• *Can you show me half of eight? Can you write the answer?*
• *Can you make 10p with your coins?*
• Do children arrange the coins in pairs (so they can see the halves more easily)?
• Do they remember to use the 5p pieces?
• *Can you show me half of 10p? Can you write the answer?*
• *How much is double five? Double five is the same as five plus five. Can you write the answer?*
• *Can you halve 12p and write down the answer? Can you double your answer and write it down?*

Task 7 Understanding doubling and halving

	AF1 – Using and applying Problem solving	AF2 – Number Operations, relationships between them
Children working at L2	**• Select the mathematics they use in some classroom activities, for example, with support.** Children should be able to draw on classroom activities with coins as well as the visual image of fingers to support finding doubles and halves as a model for the relationship between the two operations.	**• Understanding halving as a way of 'undoing' doubling and vice versa.** Children will be able to halve and double up to 10; they will know doubles up to 6 + 6 and 10 + 10, and associated halves. Children should be able to derive 7 + 7, 8 + 8, and 9 + 9 by relating to other known facts, for example, using 10 + 10 = 20 to work out 9 + 9 (i.e. using the fact that 9 is one less than 10). Children should demonstrate understanding of the relationship between doubling and halving, knowing that if double 6 is 12 then half of 12 is 6.
Children working at L3	**• Select the mathematics they use in a wider range of classroom activities.** Children should be able to solve the problem of doubling and halving a range of numbers by drawing on a range of mental calculation knowledge. Children will have good abstract understanding of the relationship between the operations.	**• Derive associated division facts from known multiplication facts.** Children will be able to double and halve tens to 50 + 50 = 100 and be familiar with near doubles such as 6 + 5 and 60 + 50 as they will be able to work out new number facts using existing knowledge. Children will be able to use strategic methods for deriving further doubles and use the relationship between them. They will thus be able to work out that the answer to *what is half of 46?* is the same as the number you double to get 46.

Task 8 Missing numbers

Framework objectives

• Understand that subtraction is the inverse of addition, using missing number sentences.
• Understand that division is the inverse of multiplication and vice versa, and use that understanding to derive related multiplication and division number sentences.

Assessment foci

AF1 – Using and applying (reasoning)
AF2 – Number (operations, relationships between them)

Resources

• Pupil Sheet 8.1 for extension activity, one per child

Task overview

Children use a range of strategies to find the missing numbers in number sentences, showing their understanding of relationships between operations.

Activity

• Write a number sentence on the board, such as 30 + ☐ = 100. Ask children to read the sentence out – for example, *30 add something equals 100;* or, *30 and how many more make 100?* Discuss strategies for solving the problem.

• Now write on the board 100 – ☐ = 30 and read the number sentence as *100 take away something equals 30*. Discuss children's strategies for solving this. Then write these problems for children to work out: ☐ + 20 = 90; 20 + ☐ = 80; 50 – ☐ = 5 and 70 – ☐ = 50.

Extension

• Use more difficult addition and subtraction problems involving multiples of ten, for example, 43 + ☐ = 83, 94 – ☐ = 54, 96 + ☐ = 126 and 87 – ☐ = 27.

• Also try this with two missing numbers. For example, ask children to write two different calculations for these problems: ☐ + ☐ = 38 and ☐ – ☐ = 54.

• Ask children to work through Pupil Sheet 8.1 to solve similar missing number problems with multiplication and division.

Observe and ask

• *How can we find out what to put in the square?*
• Do children count on with fingers up to 100?
• Do children hold up ten fingers (each digit representing 10) and fold down three (30)?
• Do children hold 100 in their head and count back 30?
• *How can we solve the subtraction problem?*
• Do children put 100 in their head and count back 30?
• *Complete these number sentences. Which problem was the easiest/most difficult? Why?*

Task 8 Missing numbers

	AF1 – Using and applying	AF2 – Number
	Reasoning	Operations, relationships between them
Children working at L2	• **Explain why an answer is correct, for example, with support.** Children will be able to explain their answers by reference to place-value understanding and by implicit understanding of the addition and subtraction relationship. With prompting they may begin to be able to explain how these are 'opposites'.	• **Use the knowledge that subtraction is the inverse of addition.** Children will be able to solve the missing number problems by using number bond knowledge (for example, since $3 + 7 = 10$, $30 + 70 = 100$). Children will make the connection between addition and subtraction sentences (for example, since $30 + 70 = 100$, $100 - 70 = 30$). By doing this, children will be implicitly showing understanding of the relationship between addition and subtraction.
Children working at L3	• **Review their work and reasoning.** Children will use their good understanding of inverse operations to explain how they have solved problems and use these to check solutions.	• **Derive associated division facts from multiplication facts.** Children will also be able to find the missing numbers using a variety of different strategies, explicitly knowing the connection between addition and subtraction. Children will be able to use the relationship between multiplication and division to solve the problems in the extension activity.

Task 9 Adding three numbers

Framework objectives
• Recognise that addition can be done in any order.

Assessment foci
AF1 – Using and applying (communicating)
AF2 – Number (mental methods)

Task overview
Children explore different strategies for adding three numbers, including 2-digit ones, discussing their methods.

Resources
• Pupil Sheet 9.1 for extension activity, one per child
• Three large 1–6 spotted dice
• Two 1–6 numbered dice (cover 1, 2 and 3 on one of the dice with 7, 8 and 9)
• 1–20 number cards

Activity
• Explain to the children that you are going to throw three spotted dice and that you want them to write down the total amount of the three sets of spots. Point out that you will be asking them to explain the strategy they used for totalling.

• After doing this several times, switch to using two numbered dice and observe and discuss the strategies children use. Use one 1–6 and one 4–9 numbered die, and a set of number cards 1–20. Throw the two dice and select a number card at random. Add up the numbers (for example, 13, 8, 5). Discuss strategies, such as: 13 + 8 = 21 (because 13 + 7 = 20, add 1 = 21), 21 + 5 = 26 (1 + 5 = 6 and 20 + 6 = 36).

Extension
• Using Pupil Sheet 9.1, ask children to add three numbers, including two larger than 20, and discuss their strategies.

Observe and ask
• *How many dots can you see (for example, six, five and four)?*
• *What strategy did you use? Can you add up the spots and give a total? For example, 6 + 4 = 10 and 5 more makes 15.*
• Did anyone use a different method to add the dice spots? For example, *Double 6 is 12, take away 1 is 11, add 4 is 15; or, Double 5 is 10, add 1 is 11, add 4 is 15; or, I put 6 in my head, counted on 5 and then counted on 4 more.*
• *What strategies did you use to add up the three numbers on the dice? Did you use any different strategies?*

Task 9 Adding three numbers

	AF1 – Using and applying Communicating	AF2 – Number Mental methods
Children working at L2	• **Discuss their work using mathematical language, for example, with support.** Children should show awareness of the variety of strategies available to them and be able to discuss these using appropriate vocabulary.	• **Use mental recall of addition and subtraction facts to 10.** Children will use their knowledge of number bonds to answer, and so switching to numbered dice should not be a problem. Most children will be comfortable working with number bonds quite quickly, and they can then be encouraged to use doubling or trebling a number as an alternative strategy.
Children working at L3	• **Discuss their mathematical work and begin to explain their thinking.** Children should be able to talk confidently about their strategies for adding three numbers and be quite systematic in choosing efficient methods based upon an analysis of the numbers, including good use of place-value understanding.	• **Add and subtract 2-digit numbers mentally.** Children will be able to total three numbers, demonstrating a range of knowledge of known facts. If they appear to take easy options when choosing the three numbers from the circles, for example, $40 + 50 + 5$, then check with more challenging examples, for example, $37 + 8 + 29$.

Task 10 Mental strategies

Framework objectives

• Rehearse addition and subtraction facts for pairs of numbers that total up to 10.
• Begin to add three 1-digit numbers mentally.

Task overview

Children have the opportunity to demonstrate knowledge of addition and subtraction bonds and to use them in different contexts.

Assessment foci

AF1 – Using and applying (reasoning)
AF2 – Number (mental methods)

Resources

• Pupil Sheet 10.1 for extension activity, one per child
• 1p coins
• A tin
• Several sets of 1–9 number cards

Activity

• Revise addition pairs for 10 – count out ten 1p coins and ask children to match the count with their fingers. Then drop some coins into a tin one at a time, without letting children see. Children listen carefully to the sound of the coins and count them quietly, perhaps using their fingers. They then write down the number of coins left.

• On the board, write 5 + ☐ = 10. Ask children to write and answer the number sentence by counting on using their fingers. Repeat a few times, dropping different numbers of coins into the tin.

• Give each child three 1–9 number cards (two of which must add up to 10, for example, 4, 5 and 6 or 7, 4 and 3) and ask them to add the three numbers. Check addition strategies – do children look for pairs to 10? Repeat with at least two more sets of cards, making sure the trios are mixed up (i.e. the pairs that add up to 10 are not adjacent).

Extension

• Give children a number sentence with two unknowns, for example:
☐ + ☐ + 6 = 16. Read out the number sentence: *Something and something add six makes 16*; or, *Six and something and something makes 16. What could the 'something' be?* Children should then independently work through the sentences on Pupil Sheet 10.1.

Observe and ask

• *How many coins are in the tin? How do you know?*
• Are children able to explain how they found the answer?
• Can children read the number sentence on the board?
• *Can you show me how you added the three numbers?*
• When adding three numbers, do children look for pairs that make 10?
• *Can you show me the two cards that make 10?*
• *Can you add on the extra number? What do you get?*
• Do children know the answer or do they count on the extra number?
• *Can you write the number sentence?*

Task 10 Mental strategies

	AF1 – Using and applying Reasoning	AF2 – Number Mental methods
Children working at L2	• **Explain why an answer is correct, for example, with support.** Children should be able to explain that adding two numbers to make 10 is helpful because they can then use place-value understanding to add the third number. They are likely to do this with the cards supporting their explanation, rather than through any abstract understanding.	• **Use mental recall of addition and subtraction facts to 10; use mental calculation strategies to solve number problems including those involving money.** Children should know their number pairs to 10. When given three cards, look to see if children immediately identify the pairs that make 10, demonstrating this and its usefulness in adding the three numbers. Some may need a little prompting to use this strategy.
Children working at L3	• **Review their work and reasoning.** Children will appreciate that different solutions are possible to the extension questions and draw upon their good number knowledge and calculation strategies to check and explain their work.	• **Add and subtract 2-digit numbers mentally.** Children will also use their knowledge of number facts and mental strategies to solve the problems in the extension activity. Children may subtract the given number from the total and then look for a pair of numbers that make up the difference. Or they may add one number then another to the given number, using place-value understanding to find easy solutions. For example, they might add a single-digit number to the given number to get up to the next multiple of ten.

Task 11 Subtracting 10s and near 10s

Framework objectives
• Subtract a multiple of 10 from a 2-digit number by counting back in 10s.
• Add and subtract 9 and 11 by adding and subtracting 10.

Task overview
Children will have opportunity to demonstrate understanding of strategies for subtracting 10s and near 10s from a range of numbers.

Assessment foci
AF1 – Using and applying (reasoning)
AF2 – Number (mental methods)

Resources
• Pupil Sheet 11.1 for extension activity, one per child
• 0–99 number square
• Tens and units place-value cards

Activity
• Practise counting in 10s on the number square. Point to 97. Ask children how many tens and how many units it has. Ask: *If we count back one ten, where will we be?* (87) *How do you know?* Check, then repeat subtracting 10 from other 2-digit numbers. Make sure that children can all explain what will happen when you subtract 10 – that the units will stay the same but the tens number will always be one fewer.

• Next practise subtracting 11 from a 2-digit number, for example, 29 – 11. Check the strategies that children use (taking away 10 then 1) before moving on to taking away 9 from a 2-digit number. Again, clarify the strategies children use to do this (taking away 10 then adding 1).Use tens and units place-value cards to add and subtract 9 and 11. Extend to adding or subtracting 19, 21, 18, 22 and near multiples of 10.

Extension
• Children should work through Pupil Sheet 11.1 to subtract near multiples of 10s from a range of 2- and 3-digit numbers, explaining their strategies.

Observe and ask
• *What happens to the tens/units numbers in 97 when we take away 10?*
• *Why does the units number stay the same?*
• *What happens if we take away 11?*
• *Is 11 more or less than 10?*
• *How many 10s are there in 11? How many units?*
• Can children write/say answers immediately?
• Do children use the number square for reference?

Task 11 Subtracting 10s and near 10s

	AF1 – Using and applying Reasoning	AF2 – Number Mental methods
Children working at L2	• **Explain why an answer is correct, for example, with support.** Children should be able to discuss and explain their strategies for subtracting near 10s, referring to number square for support. Some children may find explaining the strategy for 9 trickier (since it involves subtraction and addition rather than two subtractions).	• **Use mental recall of addition and subtraction facts to 10; use mental calculation strategies to solve number problems.** Children will be able to take away 10 from any 2-digit number. They should also be able to add and take away 11 and 9 and understand this as a two-step operation, i.e. taking away 10 then adjusting. They may be able to extend this to other near multiples of 10, for example, 19 or 21.
Children working at L3	• **Review their work and reasoning.** Children will be able to articulate their strategies for solving these questions, demonstrating good understanding of place value, referring to which digits change. They will have an equal grasp of adjusting where a subtraction and an addition are necessary.	• **Add and subtract 2-digit numbers mentally.** Children will have a good grasp of mental strategies needed to subtract a range of near multiples of 10 from a range of 2- and 3- digit numbers, involving subtracting the tens number then adjusting.

Task 12 Multiples of 2, 5 and 10

Framework objectives
• Recognise 2-digit multiples of 2, 5 and 10, and know their multiplication facts.

Assessment foci
AF1 – Using and applying (reasoning)
AF2 – Number (solving numerical problems; mental methods)

Task overview
Children are given the opportunity to demonstrate their knowledge of various multiplication facts using a counting stick, cubes and number cards.

Resources
• 10-point counting stick
• 50 cubes
• 1–50 number cards

Activity
• Rehearse counting in 2s using the 10-point counting stick with ends of 0 and 20. Stress the mid-point, i.e. 10. Choose children to write the multiples of 2 in sequence and discuss the patterns in the digits. Repeat this process for multiples of 10, and then of 5.

• Give children 50 cubes to count. Ask them to count the cubes in groups of two, five or ten, observing how they approach the task: do they count cubes individually or in groups of the designated number?

Extension
• Use number cards 1 to 20, 25, 30, 35, 40, 45 and 50. Shuffle them and reveal them one at a time. For each number ask: *Is it a multiple of 10? 5? 2?* If the answer to any of these questions is yes, challenge children to write a number sentence. For example, for card 18, they could write $9 \times 2 = 18$.

• Use this activity to discuss how some numbers (for example, 10) are multiples of 2, of 5 and of 10.

Observe and ask
• Start with multiples of 2 and point along the counting stick in ascending sequence. *Where am I pointing?*
• Point along the counting stick in descending sequence and ask *Where am I pointing?*
• Now point at different positions randomly, for example, at 6×2, and ask where you are pointing. Emphasise that this position is the sixth multiple – *six twos are twelve* and write $6 \times 2 = 12$. Repeat this process for multiples of 10 and of 5.

Task 12 Multiples of 2, 5 and 10

	AF1 – Using and applying Reasoning	AF2 – Number Solving numerical problems (L2) Mental methods (L3)
Children working at L2	• **Explain why an answer is correct, for example, with support.** Children will be able to explain why multiplication facts are correct (for example, $6 \times 2 = 12$) by reference to counting on with the counting stick to the sixth place or by counting six lots of two cubes.	• **Choose the appropriate operation when solving addition and subtraction problems.** Children will be familiar with, and able independently to count in 2s, 5s and 10s, and be able to respond to solve problems with multiplication (for example, what is 6×2) by repeated addition.
Children working at L3	• **Review their work and reasoning.** Children will be able to check work using knowledge of number facts and operations to check answers are correct.	• **Use mental recall of the 2, 5 and 10 multiplication tables.** Children will be able to use their mental recall of multiplication facts to solve these problems. Children will also use this knowledge to recognise multiples of 2, 5 and 10 beyond the tenth multiple.

Task 13 Strategies for doubling and halving

Framework objectives

- Double multiples of 5 up to 50.
- Begin to halve multiples of 10 up to 100.

Task overview

Children use a variety of mental strategies for doubling and halving a range of numbers.

Assessment foci

AF1 – Using and applying (problem solving)
AF2 – Number (mental methods; solving numerical problems)

Resources

- 5p and 10p coins

Activity

- Place two 5p coins slightly apart on the table and ask children to say *5p + 5p = 10p* and *double 5p is 10p*. Write 'double 5p = 10p' and '5p + 5p = 10p' on the board, then ask: *What is half of 10p?*

- Now put two 10p coins on the table, slightly apart, and ask: *What is double 10p? How would you write that as a number sentence? What is half of 10p?* Write '$\frac{1}{2}$ of 10p = 5p' on the board. Give each child 5p and 10p coins and repeat the activity, this time with them modelling the calculations you give, for example, doubles of 25p, 35p and 45p.

Extension

- Ask children to double and halve bigger numbers – but be systematic with your questioning. For example: *What is double 45? Double the tens first: 40 + 40 = 80, then double the units: 5 + 5 = 10, and now add them together: 8 + 10 = 90.* To halve multiples of ten, help children follow this same method. For example: *What is half of 90? Find half of 100 (50). Then find half of 80 (40). So half of 90 is 45.*

Observe and ask

- *Can you show me double 15?* Do children count out two sets of 15p? Do they write down the answer immediately? Do they count on from 15p? Do they count in fives from the beginning?
- *Can you write the answer for half of 20p?*
- *What about double/half of 50/100?*

Task 13 Strategies for doubling and halving

	AF1 – Using and applying Problem solving	AF2 – Number Mental methods (L2) Solving numerical problems (L3)
Children working at L2	• **Select the mathematics they use in some classroom activities, for example, with support.** Children should be able to demonstrate how they can use the coins to find a range of doubles and halves.	• **Use mental calculation strategies to solve number problems including those involving money.** Children should be able to recall doubles to 10 + 10 and have awareness of other significant doubles, such as double 10, 25 and 50. Children will be aware of, but sometimes find it trickier to remember, most of the corresponding halves. Children may be able to double other multiples of 10 and find half of those multiples of 10 with an even tens digit (20, 40, 60 and 80).
Children working at L3	• **Select the mathematics they use in a wider range of classroom activities.** Children will draw on a range of mental strategies, number knowledge and understanding of relationships between operations to solve these problems.	• **Solve whole number problems including those involving multiplication.** Children will be able to double multiples of 5 and find half of all multiples of 10 using a range of mental methods and knowledge of place value.

Task 14 Addition and subtraction

Framework objectives
- Add by counting on in 1s from the larger number, crossing a multiple of 10.
- Recognise addition as counting on.
- Add three numbers by putting the largest number first.
- Recognise that addition can be done in any order.
- Use the + and = signs to record addition sentences.

Task overview
Children rehearse strategies for adding three numbers, including the use of subtraction, to find missing numbers.

Assessment foci
AF1 – Using and applying (communicating)
AF2 – Number (solving numerical problems)

Resources
- Pupil Sheet 14.1, one copy per child
- Pupil Sheet 14.2 for extension activity, one copy per child

Activity
- Ask children some questions involving adding three 1-digit numbers, such as: *What is 5 and 4 and 2? What is 3 and 4 and 8?* Discuss different ways these numbers could be added together.
- Give children a copy of Pupil Sheet 14.1. Ask them to write the numbers 8, 15 and 6 in the corner circles of one of the triangles. They should add the numbers together and put the answer in the centre. They can use paper and pencil or whiteboards to record their calculations. They should put the same three numbers in the corners of a second triangle and add them a different way, working either in pairs or individually.
- Give children further sets of three numbers (choose ones that children should be able to solve) to put in the corners and again ask children to solve these questions in two different ways.
- Now ask children to put the number 20 in the centre of the triangle and numbers 7 and 5 in two of its corners. Ask children to say what number would go in the third corner. Repeat with other numbers less than 25 at the centre and at two of the three corners.

Extension
- Using Pupil Sheet 14.2, ask children to find three numbers to give the totals at the centre of the triangle and discuss their strategies.

Observe and ask
- *What would be a good way to add those numbers?*
- *Can you add them in another way?*
- *What number is missing? How can you find out what it is?*
- Do children choose efficient methods to find the missing number?
- Can children explain their choice of strategies?

Task 14 Addition and subtraction

	AF1 – Using and applying Communicating	AF2 – Number Solving numerical problems
Children working at L2	**• Discuss their work using mathematical language, for example, with support.** Children should be able to explain their strategies and, with prompting, use jottings to support this. They should be aware that addition can be done in any order. Children may need support in articulating the fact that three numbers can be added in any order and applying this idea.	**• Solve number problems.** Children will be able to add three numbers together using a range of calculation strategies (for example, putting the largest number first). Children will use 10 and other decade numbers to support counting on as well as other known facts (for example, they might add 8, 15 and 6 as 15 + 14 because they know that 8 and 6 are 14, and they can then use a near double). Where they do not have such strategies they should still be able to find answers accurately by using an empty number line to support them with counting on.
Children working at L3	**• Begin to organise their work and check results.** Children should demonstrate a systematic approach to the extension activity and draw on a variety of strategies. They will also be able to discuss the relative merits of different ways of solving a problem. Children should be able to find a range of solutions to the extension tasks.	**• Use mental recall of addition and subtraction facts to 20 in solving problems with larger numbers.** Children will use knowledge of known facts (for example, pairs of numbers that make 20 or 50) to find efficient solutions to the extension problems.

Task 15 Subtracting

Framework objectives
- Count back in 1s, not crossing a multiple of 10.
- Extend understanding of subtraction as taking away.

Task overview
Children use mental methods to subtract 1- and 2-digit numbers across tens boundaries.

Assessment foci
AF1 – Using and applying (communicating)
AF2 – Number (mental methods)

Resources
- 10p and 1p coins

Activity

- Write a subtraction on the board involving taking a 1-digit number from a 2-digit number, such as 24 − 4 = ☐. Ask children: *What have I written?* Encourage as many different responses as possible. For example: *24 take away four; 24 subtract four; make 24 four smaller; count back four from 24.*

- Then ask: *What is the answer? How did you work it out?* (Note that children who are still counting back will find this assessment difficult and so you may like to model the subtraction with 10p and 1p coins.)

- Now write 24 − 6 on the board and ask: *How many to get back to 20?* (4) *How many more to take away?* (2) *What is the answer?*

- Write 34 − 4 = ☐ and 34 − 5 = ☐ on the board, and ask children to complete the number sentences. Discuss the strategies that children use.

Extension

- Give children further calculations that involve subtracting of 1- and 2-digit numbers from 2-digit numbers and crossing the tens. For example: 37 − 8; 47 − 8; 86 − 29, 72 − 38. Discuss the steps used.

Observe and ask

- *Who can read what I have written?*
- *How can you write and complete the number sentence? How did you work out the answer?*
- Do children say: *I used five fingers and I counted back;* or, *I know that 24 − 4 is 20 so 24 − 6 will be taking two more away, so that is 18*?
- *Who can answer 34 − 4? How did you work it out?*
- Do children use a range of strategies appropriate to the numbers?

Task 15 Subtracting

	AF1 – Using and applying Communicating	AF2 – Number Mental methods
Children working at L2	**• Discuss their work using mathematical language, for example, with support.** Children should, with support, be able to explain a strategic method for subtracting a single-digit number and show some awareness of how this is more efficient than simply counting backwards.	**• Use mental recall of addition and subtraction facts to 10.** Children will be able to count back in 1s, however, in this assessment you are trying to establish whether they can split the single-digit number into meaningful parts in order to get to a multiple of 10 before subtracting the remaining part. Children should be able to use mental recall of number facts rather than simple counting in order to do this, using these for different multiples of 10.
Children working at L3	**• Discuss their mathematical work and begin to explain their thinking.** Children should be able to confidently explain their strategies for solving these subtraction problems, explaining concisely the role of the multiples of 10 ('bridging the 10') in providing an efficient mental strategy.	**• Add and subtract 2-digit numbers mentally.** Children will also be able to use two-step operations to subtract larger single-digit numbers from 2-digit numbers while bridging the tens. They will use their knowledge of number bonds to split the single-digit number into two appropriate parts (firstly to get to a multiple of 10 and then to reach the answer). Children should use these strategies to support subtracting 2-digit numbers by strategies such as first subtracting the tens and then subtracting the units parts (i.e. by partitioning).

Task 16 Addition and subtraction: problems with money

Framework objectives

• Recognise all coins and begin to use £/p notation for money.
• Solve 'real-life' problems involving money (paying an exact sum).
• Solve problems involving addition, subtraction, multiplication or division in contexts of numbers, measures or pounds and pence.

Task overview

Children solve a range of addition and subtraction problems in the context of money.

Assessment foci

AF1 – Using and applying (communicating)
AF2 – Number (solving numerical problems)

Resources

• Pupil Sheet 16.1, one per child – or label eight toys with the prices on Pupil Sheet 16.1
• Selection of coins

Activity

• Check that children can identify coins and also understand how smaller and larger value coins can be exchanged; for example, explain that a 10p coin is equivalent to two 5p coins, five 2p coins or ten 1p coins. Hold up a toy with a price label or use one of the teddy bears on Pupil Sheet 16.1. Ask children to suggest different coins that could be used to pay for it.

• Now ask children to select any two of the toys/teddy bears and find the total cost. They should repeat this four times for different pairs of toys or teddy bears. Children may use the coins for support or record as number sentences as appropriate.

• Next tell children they have 50p (hold up a 50p piece) to spend on any toy/teddy. Ask how much change they would get from your 50p. Repeat four times using different toys/teddies.

Extension

• Ask children to find the total cost of three toys/teddies and find change from £1 or £2.

Observe and ask

• *What coins would you use to pay for that toy/teddy?*
• *How much would those two toys/teddies cost altogether? How can you be sure?*
• *How much change from 50p for that toy/teddy? How can you be sure?*
• Can children find total costs or change from 50p?
• Can children record/explain this correctly?

Task 16 Addition and subtraction: problems with money

	AF1 – Using and applying Communicating	AF2 – Number Solving numerical problems
Children working at L2	• **Begin to represent their work using symbols and simple diagrams, for example, with support.** Children should be able to explain how to use coins to find the total cost of two items, or the change from 50p for one item and begin to record their work in number sentences.	• **Solve number problems involving money.** Children should be able to confidently identify different coins and know which to use to buy different items. They should be able to find the total cost of any pair of toys/teddies and record the calculation in a number sentence, possibly using coins used for each of the items to help them find the total cost. Children should be able to find the change from 50p by counting on ('shopkeeper's addition') and be able to model this using actual coins.
Children working at L3	• **Begin to organise their work and check results.** Children will be able to explain how they are choosing to solve the problems; they will work in an organised way and demonstrate strategies for checking their work.	• **Use mental recall of addition and subtraction facts to 20 in solving problems involving larger numbers.** Children are likely to use partitioning strategies and number fact knowledge, as well as a range of mental calculation strategies, to find the cost of any two toys. Children will be able to use these skills to find change from £1 or £2.

Task 17 Division as the inverse of multiplication

<table>
<tr>
<td>

Framework objectives

• Record multiplication facts using × and = in number sentences.
• Introduce division as the inverse of multiplication.

</td>
<td>

Assessment foci

AF1 – Using and applying (problem solving)
AF2 – Number (solving numerical problems)

</td>
</tr>
<tr>
<td>

Task overview

Children solve various multiplication and division problems involving the 2s, 5s and 10s times-tables.

</td>
<td>

Resources

• Pupil Sheet 17.1 for extension activity, one per child
• 1–100 number square
• 10p, 5p and 1p coins
• Number fans

</td>
</tr>
</table>

Activity

• Rehearse counting in 2s, 5s and 10s, matching the count with fingers and on the number square. You could also show the calculations using 2p, 5p and 10p coins.

• Give out number fans and ask questions so children can show their answers. For example: 5 × 2, 6 × 2, 5 × 5, 6 × 5, 10 × 5 and 9 × 5. Prompt the group with questions:
  o *If five lots of two equals ten, what will four lots of two be?* (Eight – or one lot of two less than ten.) Model this with coins.
  o *Can you use this strategy to find six times two? Nine times two? Eleven times two?*

• Remind children that multiplication is repeated addition and ask them to write each number sentence using the × symbol. Then work on some inverse division calculations, for example, *What is five times ten?* (Five lots of 10p.) *So how many 10p pieces are in 50p?* Model the sum with coins and the number square.

Extension

• Using Pupil Sheet 17.1, children should solve multiplication and division problems in the context of money.

Answers:
1a 12p; **b** 18p; **c** 26p
2a 8; **b** 12; **c** 13
3a 35p; **b** 55p; **c** 75p
4a 9; **b** 13; **c** 6
5a 60p; **b** £1.20; **c** £2.10
6a 7; **b** 9; **c** 7

Observe and ask

• Can children show the counting in 2s pattern on the number square? With coins?
• *What are four lots of 2p? Show me the answer using the fans. What strategy did you use?*
• Do children use strategic means to find particular multiplication and division facts?
• Do children recall some facts?
• Do children use multiplication facts to solve division problems?

Task 17 Division as the inverse of multiplication

	AF1 – Using and applying Problem solving	AF2 – Number Solving numerical problems
Children working at L2	**• Select the mathematics they use in some classroom activities, for example, with support.** Children will, some with prompting, use a practical resource to give a range of multiplication and division facts. For example, they will carry out repeated addition using coins or a number square to support them.	**• Choose the appropriate operation when solving addition and subtraction problems.** Children will be able to count in 2s, 5s and 10s. They should know ×5 and ×10 tables for 2, 5 and 10. They should be able to work out near multiples, for example, (4 × 2, 6 × 2, 9 × 2 and 11 × 2), using the strategies practised in the activity that are essential for quick calculation. Most will be able to divide by 2 and 10 for familiar multiples and recognise the relationship between the two operations.
Children working at L3	**• Try different approaches and find ways of overcoming difficulties that arise when they are solving problems.** Children will draw upon good number knowledge, particularly of the relationship between operations in order to solve the problems on Pupil Sheet 17.1. Children will be able, some with prompting, to use inverse operations to check their answers.	**• Solve whole number problems including those involving multiplication or division that may give rise to remainders.** Children will be able to solve most of the money problems by applying knowledge of multiplication and division facts. Some may find the division questions slightly trickier when the question of a remainder arises (such as in *How many 2p lollies would you be able to buy with 27p?*)

Task 18 Addition and subtraction problems

Framework objectives
• Use the +, – and = signs to record addition and subtraction sentences.

Assessment foci
AF1 – Using and applying (communicating)
AF2 – Number (written methods)

Task overview
Children solve a range of addition and subtraction questions, with a focus on recording the operations in number sentences or using a range of written recordings.

Resources
• Pupil Sheet 18.1 for extension activity, one per child
• Cards showing +, – and = operation signs
• Jars, one per child
• Up to 25 cubes or conkers per child

Activity
• Show children the cards for the +, – and = operation signs. Check that they can read each one and know what it signifies; if they are unable to do this then they are not ready for this assessment.

• Count six cubes/conkers into a jar. Now put in three more and ask how many cubes/conkers there are altogether. Ask children to write down what has been done. For this first example, if necessary, model writing the number sentence: 6 + 3 = 9. Now take two cubes/conkers out of the jar and ask how many cubes/conkers are left and how this can be recorded. Repeat with two or three further examples.

• Now give all the children 25 cubes/conkers and a container each and ask them to make up some problems like this of their own, recording their problems in number sentences.

Extension
• Discuss with children different ways of solving the problems on Pupil Sheet 18.1, for example, using an empty number line, column methods. Then ask them to solve the problems using any written method of their choice.

Answers: 1 91; **2** 111; **3** 196; **4** 184; **5** 223; **6** 315.

Observe and ask
• *How many cubes/conkers are in the jar now? How do you know?*
• *Can you write a number sentence to show that?*
• Do children write appropriate number sentences using the operation signs correctly?
• *You have written … 4 + 3 = 7. Can you explain that with the cubes/conkers?*
• *Why have you recorded it that way?*

Task 18 Addition and subtraction problems

	AF1 – Using and applying Communicationg	AF2 – Number Written methods
Children working at L2	• **Begin to represent their work using symbols and simple diagrams, for example, with support.** Children will be able to create their own number sentences and explain the operations they have written down using the cubes/conkers and appropriate mathematical vocabulary.	• **Record their work in writing.** Children should be able to do the arithmetic involved in this activity so the assessment should focus on the recording – children should record a number sentence for each of the operations. Children may occasionally miss out one of the signs, for example, recording 5 + 3 8, or sometimes put a sign in the wrong place, for example 5 3 + = 8.
Children working at L3	• **Use and interpret mathematical symbols and diagrams.** Children will be aware of and be able to explain how different written ways of solving addition are helpful to find efficient solutions; this should include use of place-value vocabulary to explain the methods.	• **Add and subtract 3-digit numbers using written methods.** Children should be able to record the solutions to addition problems in different ways, for example, using an empty number line, or, if it has been taught, a column method (either expanded or compact).

Task 19 Recording number problems

Framework objectives
- Record number facts using +, – and =.
- Add two 2-digit numbers.
- Subtract one 2-digit number from another.

Assessment foci
AF1 – Using and applying (problem solving)

AF2 – Number (written methods)

Task overview
Children identify which operation (addition or subtraction) to use to solve some simple word problems and have the opportunity to record these in number sentences.

Resources
- Pupil Sheet 19.1 (cut in two), one per child
- Counters or cubes

Activity

- Give each child a copy of the top half of Pupil Sheet 19.1. Read through the first problem together with the children, asking them to identify which operation (addition or subtraction) is needed to solve the problem. Ask them to solve the problem using whichever method they like, then to record the operation they have carried out in a number sentence using addition/subtraction and equals symbols. Explain that you will be looking to see that they have used these correctly as well as whether they have found the correct answer.

- Children should work through the remaining questions independently, identifying the correct operation and recording their solutions in number sentences.

Extension

- Children should work through the questions on the lower half of Pupil Sheet 19.1 independently; they may choose to solve these mentally but should be asked to then record using a written method of their choice (for example, an empty number line).

Observe and ask

- *How will you solve that problem?*
- *Is it an addition or a subtraction problem?*
- *Can you write a number sentence to show what you have done?*
- *Can you find a way to check your answer?*
- Do children identify the correct operation for each question?
- Do children's recordings represent the problems correctly?

Task 19 Recording number problems

	AF1 – Using and applying Problem solving	AF2 – Number Written methods
Children working at L2	• **Select the mathematics they use in some classroom activities, for example, with support.** Children should be able to identify the information needed to solve each problem and represent it accurately, modelling with objects and abstractly in a number sentence.	• **Record their work in writing.** Children should be able to identify the correct operation in each case, although some may need prompting with question 5 on Pupil Sheet 19.1, which involves finding the difference. Children should be able to answer the questions using recall of facts, or with the support of objects, and record in an accurate number sentence with the symbols and numbers in the correct places.
Children working at L3	• **Select the mathematics they use in a wider range of classroom activities.** Children will be able to break these problems into their own words and use a range of previous experiences of methods of solving addition and subtraction problems to solve them.	• **Add and subtract numbers using written methods.** Children should be able to confidently identify how to solve these questions. They will be able to record their answers using different written methods and check these using alternative mental strategies.

Task 20 Shape properties

Framework objectives
• Use the names of common 2D shapes, including: pentagon, hexagon, octagon.
• Sort 2D shapes and describe their features: number of sides and corners.
• Classify and describe common 2D and 3D shapes.

Task overview
Children classify and sort shapes using a range of properties such as sides, angles and symmetry.

Assessment foci
AF1 – Using and applying (communicating)
AF3 – Shape, space and measures (properties of shape)

Resources
• Pupil Sheet 20.1 (ideally enlarged to A3 size), one per child
• Plastic shapes
• Scissors
• Sugar paper

Activity

• Spread some plastic shapes on the floor. In turn, ask each child to choose a shape and, without indicating which one it is, describe its properties. The other children should try to work out which one it is from the description.

• Give children Pupil Sheet 20.1 and ask them which of the shapes are like the plastic ones and which ones are different. For those that are seen as 'different', can children identify any shared characteristics? Establish that shapes can be sorted by the number of sides they have – for example, triangles, quadrilaterals (including squares and rectangles), pentagons, hexagons and octagons. Check that children know how many sides each of these shapes have.

• Ask children to cut up the sheet and stick the shapes on sugar paper under the headings 'triangles', 'quadrilaterals', 'pentagons', 'hexagons', 'octagons'.

Note: an alternative way to present the task would be to ask children to find, for each shape on the sheet, one other shape that would go in the same group, and draw a line to link them.

Extension

• Challenge children to sort shapes into regular and irregular or symmetrical/non-symmetrical categories and then draw some further pentagons, hexagons and octagons with those properties.

Observe and ask

• *How can you describe that shape?*
• *What can you tell me about that shape that is similar or different to that one?*
• *How could you sort the shapes?*
• Do children identify properties of different shapes?
• Can children sort the shapes on the sheet according to their number of sides/corners?

Task 20 Shape properties

	AF1 – Using and applying Communicating	AF3 – Shape, space and measures Properties of shape
Children working at L2	• **Discuss their work using mathematical language, for example, with support.** Children will be able, some with prompting, to use a range of appropriate mathematical vocabulary to describe shapes and be able to give examples of similarities and differences.	• **Use mathematical names for common 3D and 2D shapes; describe their properties, including numbers of sides and corners.** Children should be able to name most of the shapes on the sheet by reference to their number of sides and corners, although they may occasionally miscount the number of sides of the pentagons, hexagons and octagons. Some children may be confused by the varied orientations of some of the shapes.
Children working at L3	• **Discuss their mathematical work and begin to explain their thinking.** Children will be able to give concise explanations of the similarities and differences between shapes. Children will identify and describe a range of their features (for example, symmetry as well as the more obvious number of sides/corners). Children will be able to quickly match shapes which share the same property and explain why that is so.	• **Classify 3D and 2D shapes in various ways using mathematical properties.** Children ought not be confused by the varied orientations of some of the shapes. They will also be able to sort by further properties such as symmetrical/non-symmetrical and regular/irregular.

Task 21 3D shapes

Framework objectives
- Sort 3D shapes and describe their features: number of faces and corners.

Task overview
Children explore the properties of 3D shapes, using a range of mathematical vocabulary to demonstrate their understanding.

Assessment foci
AF1 – Using and applying (communicating)
AF3 – Shape, space and measures (properties of shape)

Resources
- 3D shapes (pyramid, cuboid, cube, cylinder, cone and sphere)
- Labels marked 'pyramid', 'cuboid', 'cube', 'cylinder', 'cone', 'sphere'
- Labels marked 'circular', 'triangular', 'rectangular', 'square faces'
- Feely bag
- Cards cut into 2D shapes (square, circular, triangular, rectangular)

Activity

- Give one of each 3D shape to children and display the labels (see resources). Read the labels together and ask each child to say at least two things to describe each of their shapes in turn. Discuss the number of faces/corners/edges/types of face on their shape. Children should find the correct 3D shape label and stick it to the shape.

- Now place the matching set of 2D cards into a feely bag. Children should take turns to feel a shape in the bag and then describes it to the group. The group tries to identify the shape by finding the matching 3D shape. Check by matching the shape and the correct label.

Extension

- Ask children to classify and discuss shapes using more complex properties such as right angles, symmetry, regular/irregular.

Observe and ask

- *How many faces can you see on your shape?*
- Do children count accurately or do they just know?
- *Can you name the faces?*
- *Does your shape have flat or curved faces?*
- *How many corners/edges does it have?*
- *Find the label to match your shape. How do you know it is that shape?*
- Can children describe the shape in the bag?
- Do children need to be prompted to describe the shape?

Task 21 3D shapes

	AF1 – Using and applying Communicating	AF3 – Shape, space and measures Properties of shape
Children working at L2	• **Discuss their work using mathematical language, for example, with support.** Children will, with vocabulary prompts, be able to use a range of mathematical language to discuss features of the various 3D shapes.	• **Use mathematical names for common 3D and 2D shapes; describe their properties, including numbers of sides and corners.** Children will be able to describe a range of properties of the 3D shapes, including faces, corners and edges. Children should be able to identify the 2D shapes that are the faces of the 3D shapes. In most cases they should be able to use these features to identify the correct names for the 3D shapes, although they may find distinguishing cubes and cuboids difficult.
Children working at L3	• **Discuss their mathematical work and begin to explain their thinking.** Children will be articulate in using a range of mathematical vocabulary to describe features of the full range of shapes.	• **Classify 3D and 2D shapes in various ways using mathematical properties.** Children will have good knowledge of the names and features of the full range of 3D shapes, identifying them quickly. Children will be able to classify the shapes using a wider range of features such as right angles, symmetry, regular/irregular.

Task 22 Position and movement: positions

Framework objectives
• Describe the positions of objects.

Task overview
Children examine a grid of shapes in different orientations and discuss their relative positions.

Assessment foci
AF1– Using and applying (communicating)
AF3 – Shape, space and measures (properties of position and movement)

Resources
• Pupil Sheet 22.1, one per child and one enlarged copy for reference
• Squared paper and pencils

Activity
• Give each child a copy of Pupil Sheet 22.1 and pin up an enlarged copy for reference. Ask children to name the shapes and to describe the similarities between each. Ask them to locate particular shapes on the grid by pointing at them. Ask questions, for example:
  ○ *Where is the rectangle with dots?*
  ○ *Where is the rectangle with the plain heart?*
  ○ *Where is the rectangle with the cross-hatched star?*
  ○ *Where is the rectangle with the grey rectangle?*

• Ask questions about the relative positions, for example:
  ○ *Which shape is one place to the left of the grey triangle?*
  ○ *Which shape is two places to the right of the black heart?*
  ○ *Which is one place above the cross-hatched rectangle?*
  ○ *Which is three places below the black star?*

• Ask children to make up similar questions. Go around the group, inviting children to ask their questions to each other.

Extension
• Children should label the bottom of the grid A–D (left to right by column) and the side 1–4 (bottom to top by row). They should then repeat the original activity using grid references.

• Now ask them to focus on each of the shapes in turn and to explain how each has been rotated from row to row. Give children squared paper and ask them to draw up a similar 4 by 4 grid and to use the capital letters A, T, H and E in each row, rotating them by 90° in each successive row.

Observe and ask
• Can children describe features of different shapes?
• Can children locate shapes with different properties on the grid?
• Can children locate shapes given the instructions?
• Can children pose similar problems using positional language?

Task 22 Position and movement: positions

	AF1 – Using and applying Communicating	AF3 – Shape, space and measures Properties of position and movement
Children working at L2	• **Discuss their work using mathematical language, for example, with support.** Children should be able to understand and use a range of positional language to answer the questions, and to pose similar ones about the relative positions of the shapes. Some children may need prompting to use the associated vocabulary accurately.	• **Describe the positions of objects.** Children should be able to recognise the same shapes in different orientations. Children should also be able to locate different shapes in their relative positions.
Children working at L3	• **Discuss their mathematical work and begin to explain their thinking.** Children should be able to use a range of vocabulary, including that of position and turn, to describe the movements and positions of the shapes.	• **Recognise shapes in different orientations.** Children should be able to identify the shifts, i.e. 90° left/right or clockwise/anti-clockwise, between the shapes on successive rows and be able to construct a similar grid using letters rather than shapes.

Task 23 Position and movement: routes

Framework objectives
- Give directions along a route.
- Describe a route using positional language.

Task overview
Children examine a diagram showing the positions of a house, church, school and shop. They describe the routes between the different buildings.

Assessment foci
AF1 – Using and applying (communicating)
AF3 – Shape, space and measures (properties of position and movement)

Resources
- Pupil Sheet 23.1, one per child
- Counters, one per child

Activity
- Check that children are familiar with terms such as left, right and forward. Give each child a copy of Pupil Sheet 23.1 and a counter. Ask them to place the counter at the school and then to move it to the house. Discuss the route taken (two routes are possible). Ask them how they could record this route (children may need some prompting).

- Ask children to record the route so that someone else can follow it, using language such as *Turn left out of the school, move forward 2 squares, turn right, move forward 6 squares, turn right …* If necessary, start this off as a group, together modelling the instructions.

- Finally ask children to choose other pairs of locations and repeat this process.

Extension
- Ask children to place two more buildings on the diagram – a police station and a town hall. They should then describe more complex routes involving stopping. For example, how to get from the church to school via the town hall.

Observe and ask
- *Can you move the counter from the house to the school?*
- *Is there another route you could take?*
- *Which direction do you turn (i.e. left or right) when you get to that corner or junction?*
- Can children describe the route between two locations?

Task 23 Position and movement: routes

	AF1 – Using and applying Communicating	AF3 – Shape, space and measures Properties of position and movement
Children working at L2	• **Discuss their work using mathematical language, for example, with support.** Children should be able to discuss this activity and describe their routes using appropriate vocabulary, although some may need prompting to use this accurately.	• **Distinguish between straight and turning movements.** Children should be able to distinguish between right and left turns (although some may need prompting to think of this as relative to the direction they are moving in). They will also give an accurate description of the route between locations on the grid.
Children working at L3	• **Discuss their mathematical work and begin to explain their thinking.** Children should be able to accurately use a range of mathematical vocabulary to describe the routes around the grid.	• **Describe position and movement.** Children should be able to give a concise and accurate set of directions for a journey between any two places on the grid.

Task 24 Measuring and comparing length using centimetres

Framework objectives
• Estimate, measure and compare lengths using standard units (centimetres).

Task overview
Children make models of different lengths and measure lines to the nearest centimetre and half centimetre.

Assessment foci
AF1 – Using and applying (problem solving)
AF3 – Shape, space and measures (measures)

Resources
• Pupil Sheet 24.1 for extension activity, one per child
• Salt dough
• Infant centimetre rulers
• Conventional primary rulers for extension activity

Activity
• Give each child a lump of salt dough and an infant centimetre ruler. Take time to look at the ruler and establish that each demarcation is one centimetre, that each centimetre is exactly the same length and that it is the demarcation *after* the number and not the space in between that is used to establish the length. Ask the children to make one salt-dough snake 15 cm long and another 9 cm long.

• Children can then make more snakes with a difference of 4 cm, before moving on to make three or more snakes, each with a difference of 3 cm.

Extension
• Using Pupil Sheet 24.1, children measure lengths and draw lengths to the sizes given.

Answers for section 1: A 12 cm; **B** 19 cm; **C** $5\frac{1}{2}$ cm; **D** $17\frac{1}{2}$ cm; **E** 13 cm; **F** $12\frac{1}{2}$ cm.

Observe and ask
• *How long is your ruler?*
• *What units do you use the ruler to measure in?*
• *Are all the centimetres the same length?*
• *Can you point to where you will begin measuring?* (Check that children are pointing to the base line.)
• *If your snake was 6 centimetres long, show me where it would start and finish on your ruler.*
• *Make a 15 cm-long snake and a 9 cm-long snake. Which is the longer? How much longer? How do you know?*

Task 24 Measuring and comparing length using centimetres

	AF1 – Using and applying Problem solving	AF3 – Shape, space and measures Measures
Children working at L2	• **Select the mathematics they use in some classroom activities, for example, with support.** Children should draw on previous experiences of using a ruler when approaching these problems. Where they begin making errors such as those described on the right, correct them and see if these are then avoided once they are pointed out. Children should draw on subtraction knowledge to say which snake is longer and which is shorter and, with prompting, will be able to find the difference.	• **Begin to use everyday non-standard and standard units to measure length and mass.** Children will be able to make the snakes and measure them accurately. They may initially make the snakes by estimating and be able to adjust the length using the ruler. Some may measure up to the number on the ruler (to a halfway number) but not up to the demarcation after the number. Some will measure to the demarcation before the number and some will not align the snake with the base line. If they make any of the above errors, prompt children to see if they can correct them. Some may find the problem of making two snakes with a specific difference in length more challenging.
Children working at L3	• **Try different approaches and find ways of overcoming difficulties that arise when they are solving problems.** Children will appreciate that a variety of solutions are possible to the problem of making two snakes with a specific difference in length. Children will draw upon previous experiences of using a ruler in order to efficiently measure and draw lines, including examples of lines in different orientations.	• **Use non-standard units and standard metric units of length, capacity and mass in a range of contexts.** Children will be able to successfully make snakes of a given difference in length. They will be able to complete the tasks on Pupil Sheet 24.1, efficiently using a ruler both to measure and draw lines to the nearest centimetre and half centimetre, showing good understanding of how to line up the zero point on the ruler (rather than the end of the ruler) with the end of the line.

Task 25 Recognising right angles

Framework objectives
- Recognise whole, half and quarter turns.
- Know that a right angle is a measure of a quarter turn.
- Recognise right angles in simple shapes.
- Recognise clockwise and anti-clockwise turns.

Task overview
Children use an angle checker to identify right angles on shapes and in some more complex drawings.

Assessment foci
AF1 – Using and applying (problem solving)
AF3 – Shape, space and measures (measures)

Resources
- Pupil Sheet 25.1, one per child
- Plastic shapes
- Marker pens
- Rulers, paper and pencils

Activity
- Show children some plastic shapes (for example, squares, triangles, pentagons, rectangles). Check that they know what a right angle is and which of the shapes in front of them have one or more right angles. You could sort the shapes to show which have right angles and which do not.

- Remind children how to make a right angle checker by folding a piece of paper in half and then in half again. They should use the right angle checker to find the right angles in the shapes.

- Give each child a copy of Pupil Sheet 25.1. This could be cut up so that each shape arrangement can be considered separately and, for some children, enlarged. Ask children to use their right angle checker to find all of the right angles on each shape, marking the corners where they find these with a marker.

Answers: There are 9 right angles on the top shape arrangement and 12 (including 2 external ones) on the lower shape arrangement.

Extension
- Give children a square piece of paper, ruler and pencil. Ask them to create their own shape arrangement that could be checked by a friend for right angles.

Observe and ask
- *Can you describe a right angle?*
- *Which shapes have right angles?*
- *How can you use the right angle checker?*
- Do children correctly identify the right angles in the two shape arrangements?

Task 25 Recognising right angles

	AF1 – Using and applying Problem solving	AF3 – Shape, space and measures Measures
Children working at L2	• **Select the mathematics they use in some classroom activities, for example, with support.** Children should be able to use the right angle checker and make a connection to similar situations where they have used one previously.	• **Begin to use a wider range of measures.** Children will be able to recall what a right angle looks like and which shapes have one or more. They should be able to locate most of the right angles on the shape arrangements. Children may need prompting to spot those that are not in horizontal–vertical orientations and not pick up on those that are external to the shapes in the arrangement.
Children working at L3	• **Try different approaches and find ways of overcoming difficulties that arise when they are solving problems.** Children will have a good grasp of what a right angle is, and explain how to identify one using a range of appropriate mathematical vocabulary. Children will examine the complex diagrams systematically to find the right angles.	• **Use a wider range of measures.** Children will be able to find all of the right angles, including those in shapes that are not in horizontal–vertical orientations and those that are external to the arrangements.

Task 26 Mass

Framework objectives

• Estimate, measure and compare weights using standard units: kilograms, grams.
• Use and begin to read the vocabulary related to length, mass and capacity.

Task overview

Children weigh a range of objects, both by direct comparison and then by accurate use of a range of gram weights.

Assessment foci

AF1 – Using and applying (problem solving)
AF3 – Shape, space and measures (measures)

Resources

• 1 kg, 500 g, 100 g, 50 g, 20 g, 10 g, 5 g and 1 g plastic weights
• Apples, onions and carrots in different sizes
• Two identical plastic bags
• Bucket balances
• Small pan balances
• Packets with the weights concealed (some big but light; others heavy but small)

Activity

• Show children a kilogram and other gram weights and explain that there are 1000 g in a kilogram. Pass the weights round so that children can compare them. Then place some apples in one bag and a kilogram weight in the other. Invite each child to hold a bag in either hand and to say which bag is the heavier. Check using the bucket balance. Then ask children to choose a vegetable. Challenge them to make a bag of their chosen vegetable balance with a kilogram weight.

Extension

• Give the group several packages to weigh accurately on the balance scales, using different weights to balance the packages. Then encourage children to line up their packages in order, from the lightest to the heaviest.

Observe and ask

• *Which is heavier, the kilogram weight or the 1000 g weight?*
• *Can you make the vegetables balance with the kilogram? Do both sides balance?*
• *What will you need to do to make both sides balance?*
• Do children replace their vegetables with ones of different sizes to help equalise their balance? (For example, do they replace a large carrot with a small one to reduce the overall weight of their carrots?)
• Observe carefully to see how children adjust the vegetable side to achieve a balance; *How do you know that both sides weigh the same?*

Task 26 Mass

	AF1 – Using and applying Problem solving	AF3 – Shape, space and measures Measures
Children working at L2	• **Select the mathematics they use in some classroom activities, for example, with support.** Children will draw on previous experiences of using balances and, with prompting, show some strategies for adjusting the contents of the bag to achieve a balance.	• **Begin to use everyday non-standard and standard units to measure length and mass.** Children will be able to hold a bag in either hand and say which is the heavier. They will understand how to use the pan balance to find parcels that are equal in weight (see 'Observe and ask' questions) but may get frustrated at not being able to get an exact balance easily.
Children working at L3	• **Try different approaches and find ways of overcoming difficulties that arise when they are solving problems.** Children should be able to weigh accurately and will have strategies for systematically making adjustments in order to weigh objects accurately.	• **Use non-standard and standard metric units of length, capacity and mass in a range of contexts.** Children should be able to use the pan balance systematically, working from the biggest weights to the smallest, and be able to order their packages from the lightest to the heaviest.

Task 27 Reading scales

Framework objectives
• Begin to read a simple capacity scale to the nearest labelled and unlabelled division.

Task overview
Children help to label the divisions on a representation of a container and are given the opportunity to read scales with different calibrations.

Assessment foci
AF1 – Using and applying (problem solving)
AF3 – Shape, space and measures (measures)

Resources
• Pupil Sheet 27.1 (enlarged if possible)
• Litre jug of water (for reference)
• Sticky labels

Activity

• Display an enlarged copy of Pupil Sheet 27.1 and write on the board that there are 1000 ml in one litre. Read this together with the class and explain that you are together going to draw on a scale (using sticky labels) to show millilitres. Ask children if they think 1000 division lines will fit onto the jug and why/why not. Explain that you are going to draw equal divisions onto the jug, but that you will need help with labelling them. Divide the jug into hundreds and ask the children to help you to label the divisions. Label alternate divisions (200 ml, 400 ml, 600 ml, 800 ml, 1000 ml), explaining that the numbers would be too small to read if they were all written onto the jug.

• Draw some water in the actual jug. Invite children to read the scale and say how much water has been poured in. Repeat for other amounts.

Extension

• Calibrate the jug on Pupil Sheet 27.1 into hundreds with lines, the fifties with dots and label 0, 500 and 1000 ml, so that 0, 500 and 1000 are the only numbers on the scale. Ask questions about multiples of 50, and questions that relate 250, 500 and 750 ml to quarter, half and three-quarters of a litre.

Observe and ask

• *If I was asked to put a litre of water into the jug, where would the level be?*
• *How many millilitres would that be?*
• *Can you write down how many millilitres there would be if I filled the jug to this mark? How do you know? How did you work it out?*
• Can children work out answers for other calibrated but unlabelled divisions?
• *If I half-filled the jug how much water would there be? How do you know?*
• Do children immediately point to the halfway mark between 400 ml and 600 ml?
• (Point to an imaginary halfway division between 200 ml and 300 ml.) *If it was filled to this here, how much water would there be?*

Task 27 Reading scales

	AF1 – Using and applying Problem solving	AF3 – Shape, space and measures Measures
Children working at L2	• **Select the mathematics they use in some classroom activities, for example, with support.** Children will be able to make connections between the jug itself and the diagrammatic representation of it, drawing on experiences of work with liquids and the associated vocabulary of capacity, for example, 'full', 'half full', 'empty'.	• **Begin to use a wider range of measures.** Children will be aware that litres are used to measure capacity and, through this activity, show that they understand that reading the scale of the side of a jug shows how much it contains. Children will be able to read and mark the main divisions on the scale and be able to work out some of the other divisions, for example, that the jug half full would contain 500 ml.
Children working at L3	• **Select the mathematics they use in a wider range of classroom activities.** Children will draw on a good knowledge of standard units of measure, reading of scales and knowledge of the relationship between litres and millilitres to solve these problems.	• **Use standard metric units of length, capacity and mass in a range of contexts.** Children will use a good knowledge of standard units of capacity to label the jug and answer questions about the amounts in between the demarcations, as well as fractional amounts of a litre.

Task 28 Collecting data

Framework objectives
- Sort organise and interpret information in a block graph.
- Organise and interpret information in a table.

Assessment foci
AF1 – Using and applying (problem solving)
AF4 – Handling data (processing and representing data)

Task overview
Children examine a table of information and use it to draw up a tally chart and block graph.

Resources
- Pupil Sheet 28.1, one per child
- Squared paper and pencils

Activity
- Give children a copy of Pupil Sheet 28.1 each. Explain that it shows the results of a questionnaire asking children what is their favourite sport. Ask questions about which was the most popular/least popular sport and how children can find this out.

- Draw children's attention to the tally chart at the bottom of the page and ask them to use that to draw up a tally to find out how many children preferred each sport. Discuss results and then ask children to draw a block graph on squared paper showing these results.

Answers: Football 9; Tennis 4; Rounders 6; Cricket 6; Basketball 3.

Extension
- A survey of favourite sports in three classes gave the following results: Football 26; Tennis 15; Rounders 19; Cricket 14; Basketball 11. Children should use these results to draw up a bar graph in which one square represents two children.

- They should then write some sentences to explain what it shows, such as the favourite/least favourite sport. Ask: *Are the answers similar to the ones in the first activity? How many more children like football than tennis?*

Observe and ask
- *How can you find out which was the most popular/least popular sport?*
- *Did any sports get the same number of votes?*
- Can children successfully draw up a tally chart?
- Can children successfully draw up a block graph?
- Can children interpret information shown on the tally chart and block graph?

Task 28 Collecting data

	AF1 – Using and applying Problem solving	AF4 – Handling data Processing and representing data
Children working at L2	• **Select the mathematics they use in some classroom activities, for example, with support.** Children should be able to identify the information from the table needed to draw up the tally chart, although some may need prompting to do this in a systematic way. Children will draw upon previous experiences of interpreting and drawing graphs to complete the task.	• **Record results in simple lists, tables, pictograms and block graphs.** Children should be able to complete a tally chart by checking through the list of children's preferences. Children should understand the 'strike through' method of recording tallies in 5s. They should be able to use this information to draw up a block or bar graph, although in some cases they may need support with drawing up the axes and labelling them.
Children working at L3	• **Select the mathematics they use in a wider range of classroom activities.** Children will draw upon previous experiences of interpreting graphs with scales, and knowledge of multiples of 2 to successfully complete the task and be able to interpret their findings.	• **Construct bar charts, where the symbol represents a group of units.** Children will be able to independently construct an accurate bar chart from the information, labelled correctly, where one square represents two children, including half squares for odd numbers. Children should be able to explain what this shows.

Task 29 Sorting and classifying

Framework objectives	Assessment foci
• Sort and classify using different criteria. • Recognise odd and even numbers to at least 50.	AF1 – Using and applying (reasoning) AF4 – Handling data (processing and representing data)

Task overview	Resources
Children sort numbers according to criteria such as odd/even or multiples/not multiples of 5.	• Pupil Sheet 29.1 for extension activity, one per child • 1–50 number cards • Large piece of sugar paper

Activity

• Discuss the characteristics of odd and even numbers. Although this is not the main criteria for the assessment, it is important that children have that understanding. On a large sheet of sugar paper, write the headings 'odd numbers' and 'even numbers', splitting the sugar paper in two. Shuffle the 1–20 number cards and give one to each child. Ask them to place their cards on the sugar paper in the correct section. Check that all are in the right place and discuss any that are not. Choose two cards yourself and place one correctly and one incorrectly. Ask children to identify which has been placed correctly. Distribute the rest of the cards, i.e. 21–50 and ask children to sort them into the right section.

• Repeat the activity, asking children to identify numbers that are a multiple/ not a multiple of 5. Can they find any numbers that match both criteria (i.e. both odd and a multiple of 5)?

Extension

• Use the Venn diagram on Pupil Sheet 29.1. One ring represents odd numbers, the other number represents multiples of 5. Ask children to write all of the numbers 1–50 in the right sections. Ask: *Which numbers go in the overlap? Are there any numbers that do not fit in either of the rings?*

• As an alternative, children could carry out the same task using a Carroll diagram.

Observe and ask

• *How can you tell which numbers are odd, which are even?*
• *Is that number in the right place? Why/why not?*
• Can children place numbers correctly?
• Can children spot errors in placement and explain why?

Task 29 Sorting and classifying

	AF1 – Using and applying Reasoning	AF4 – Handling data Processing and representing data
Children working at L2	• **Explain why an answer is correct, for example, with support.** Children should be able to explain the placement of different number cards on the sugar paper by reference to their properties.	• **Sort objects and classify them using more than one criterion.** Children should be able to successfully sort numbers into odd and even ones and multiples of 5/not multiples of 5. Children will be beginning to be able to identify numbers with both characteristics.
Children working at L3	• **Review their work and reasoning.** Children will be able to refer to the completed Venn diagram to explain the properties of different numbers. For example: *5 is an odd multiple of 5; 17 is odd but not a multiple of 5; 30 is a multiple of 5 but not odd.*	• **Use Venn and Carroll diagrams to record their sorting and classifying of information.** Children will be able to successfully place all numbers from 1–50 on a Venn diagram. They will identify those that are both odd and multiples of 5 (i.e. 5, 15, 25 and 35) in the intersection, as well as those that fall outside the two rings of the diagram.

Task 30 Interpreting a block graph

Framework objectives	Assessment foci
• Sort, organise and interpret information in a block graph.	AF1 – Using and applying (communicating) AF4 – Handling data (interpreting data)

Task overview	Resources
Children identify information on a block graph answering a range of questions to demonstrate their understanding.	• Pupil Sheet 30.1, one copy per child • Pupil Sheet 30.2 for extension activity, one per child

Activity

• Discuss the general features of a block graph with reference to Pupil Sheet 30.1, which shows the favourite fruits of a particular class. Go on to discuss what the block graph on Pupil Sheet 30.1 shows. Ask:
   ○ *How does the height of the blocks relate to numbers of children?*
   ○ *What is the significance of the height of the block?*
   ○ *What does it mean if two blocks are the same?*

• Ask some further questions in relation to the block graph; some have been outlined on the pupil sheet. Children who are confident should be given time to work through the questions independently, noting the answers on the sheet.

Note: if you ask children to work individually, check first that they understand the vocabulary used on the pupil sheet. Other children may benefit from having the questions posed verbally and writing answers on a whiteboard, checking through each before continuing. Since this is an assessment of children's interpretation, choose the method of presentation to suit the child.

Extension

• Use Pupil Sheet 30.2. This graph has information about a similar survey with a larger group, hence the labelling in twos. Children should make up similar questions as with Pupil Sheet 30.1.

Observe and ask

• *What does the graph show?*
• *What does the height of the block tell you?*
• *How can you find the difference between the height of two of the blocks?*
• Can children use the graph to elicit the information needed to answer the questions?

Task 30 Interpreting a block graph

	AF1 – Using and applying Communicating	AF4 – Handling data Interpreting data
Children working at L2	• **Discuss their work using mathematical language, for example, with support.** Children will be able to extract the information needed to answer the questions, sometimes with a little prompting, and explain how the information from the graph supports their answers.	• **Communicate their findings, using the simple lists, tables, pictograms and block graphs they have recorded.** Children will understand the features of a block graph. Some may find the comparison questions trickier to understand than the addition ones, particularly if trying to link them to number sentences. Children will find it simpler to count on from the shorter block to the taller block, whichever way around the question is posed.
Children working at L3	• **Use and interpret mathematical symbols and diagrams.** Children will give an articulate explanation of the features of the graphs. They will answer the questions concisely, using a range of appropriate mathematical language to describe what they are doing.	• **Extract and interpret information presented in simple tables, lists, bar charts and pictograms.** Children will have a good understanding of the features of a block graph, easily extracting information to answer the questions. Children will have a clear understanding of the issue of the scale of the graph in the extension activity.

Counting on and back in 1s and 10s

1. Find the number 10 more than:

   a) 93

   b) 147

   c) 196

   d) 289

2. Find the number 20 more than:

   a) 88

   b) 134

   c) 247

   d) 392

3. Find the number 10 less than:

   a) 78

   b) 106

   c) 277

   d) 303

4. Find the number 20 less than:

   a) 114

   b) 268

   c) 305

   d) 419

Compare numbers

Put these sets of numbers in order of size, smallest to largest.

1. 204, 240, 56, 65, 378, 387

    _____ _____ _____ _____ _____ _____

2. 17, 77, 70, 7, 71, 74

    _____ _____ _____ _____ _____ _____

3. 236, 632, 362, 623, 263, 326

    _____ _____ _____ _____ _____ _____

4. 550, 515, 555, 505, 503, 305

    _____ _____ _____ _____ _____ _____

Missing numbers

Find the missing numbers in these number sentences.

1. $7 \times \boxed{} = 14$

2. $6 \times \boxed{} = 30$

3. $\boxed{} \times 10 = 40$

4. $\boxed{} \times 7 = 21$

5. $8 \times \boxed{} = 32$

6. $16 \div \boxed{} = 8$

7. $45 \div \boxed{} = 9$

8. $80 \div \boxed{} = 10$

9. $24 \div \boxed{} = 6$

10. $21 \div \boxed{} = 7$

Adding three numbers

Choose one number from each circle. Use mental methods to add them up. Record your answers in number sentences.

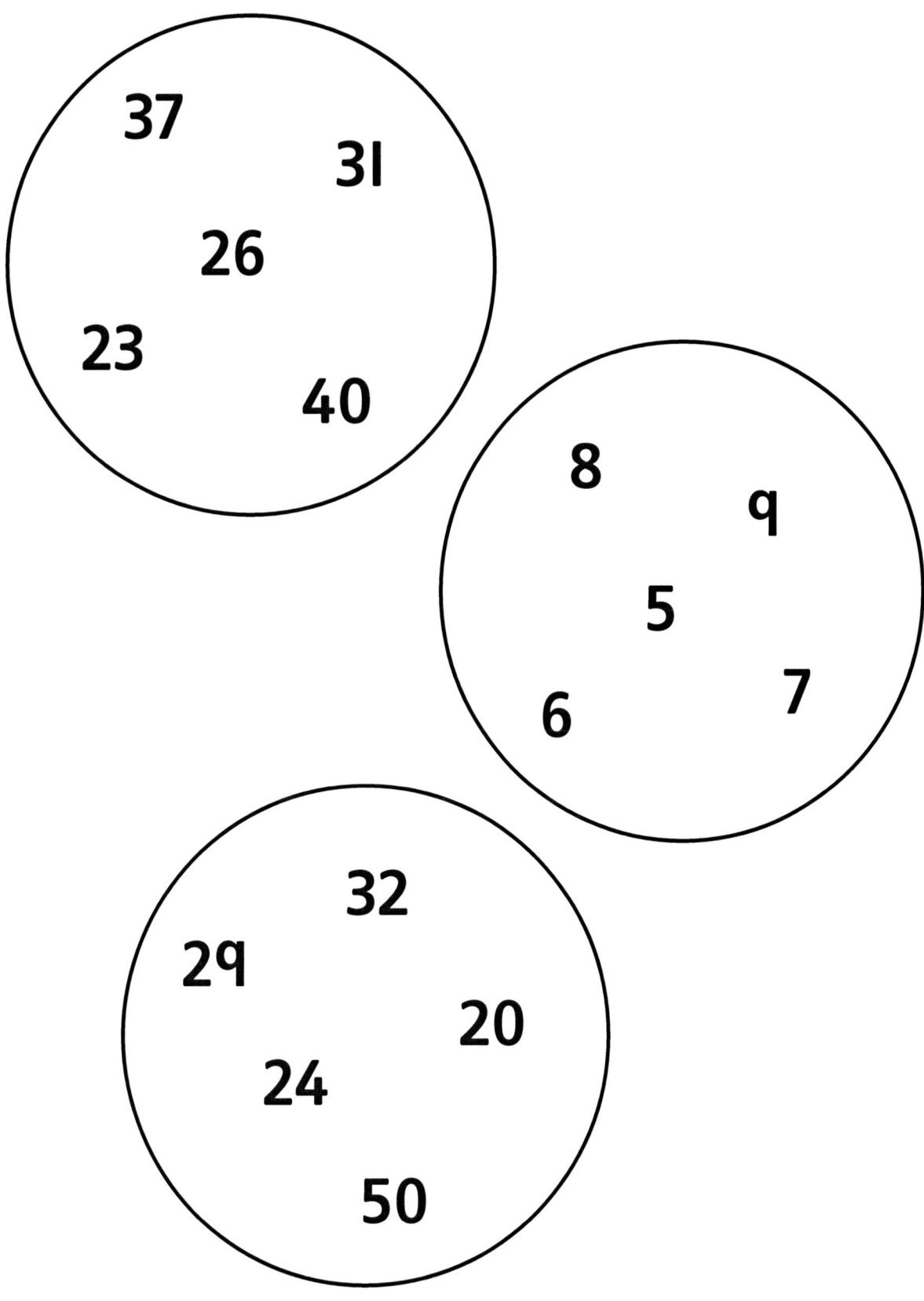

Mental strategies

What could the missing numbers be in each of these number sentences? Find at least three answers for each question.

1. ☐ + ☐ + 6 = 16

2. ☐ + ☐ + 9 = 25

3. ☐ + 17 ☐ + ☐ = 38

4. 19 ☐ + ☐ + ☐ = 50

5. ☐ + ☐ + 23 = 61

Subtracting

Choose one number from each circle. Find the difference between the two numbers. Repeat at least 10 times. Use different numbers each time.

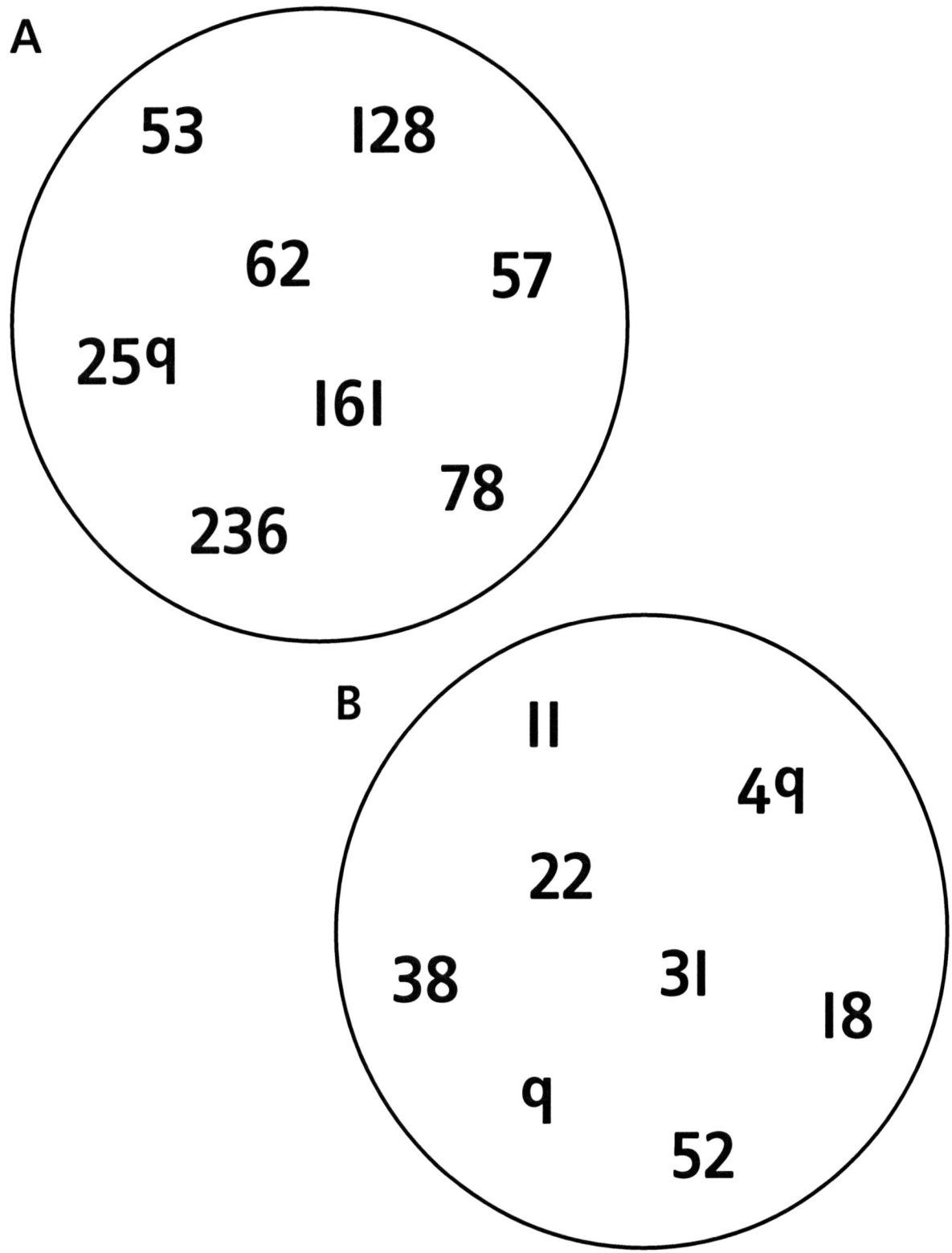

A

53   128

62   57

259

161

78

236

B

11

49

22

38   31

18

9

52

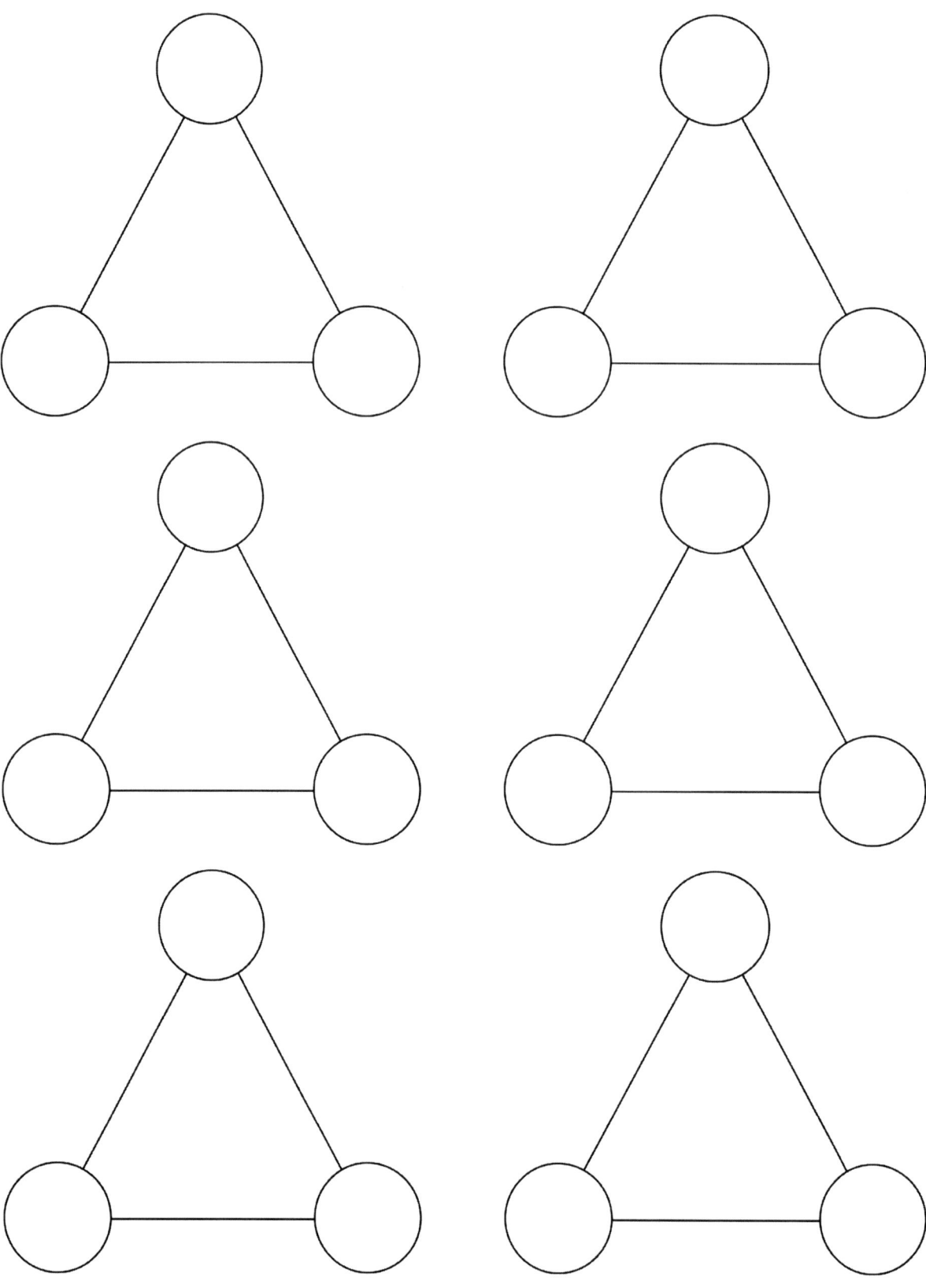

Adding three single-digit numbers

Addition and subtraction

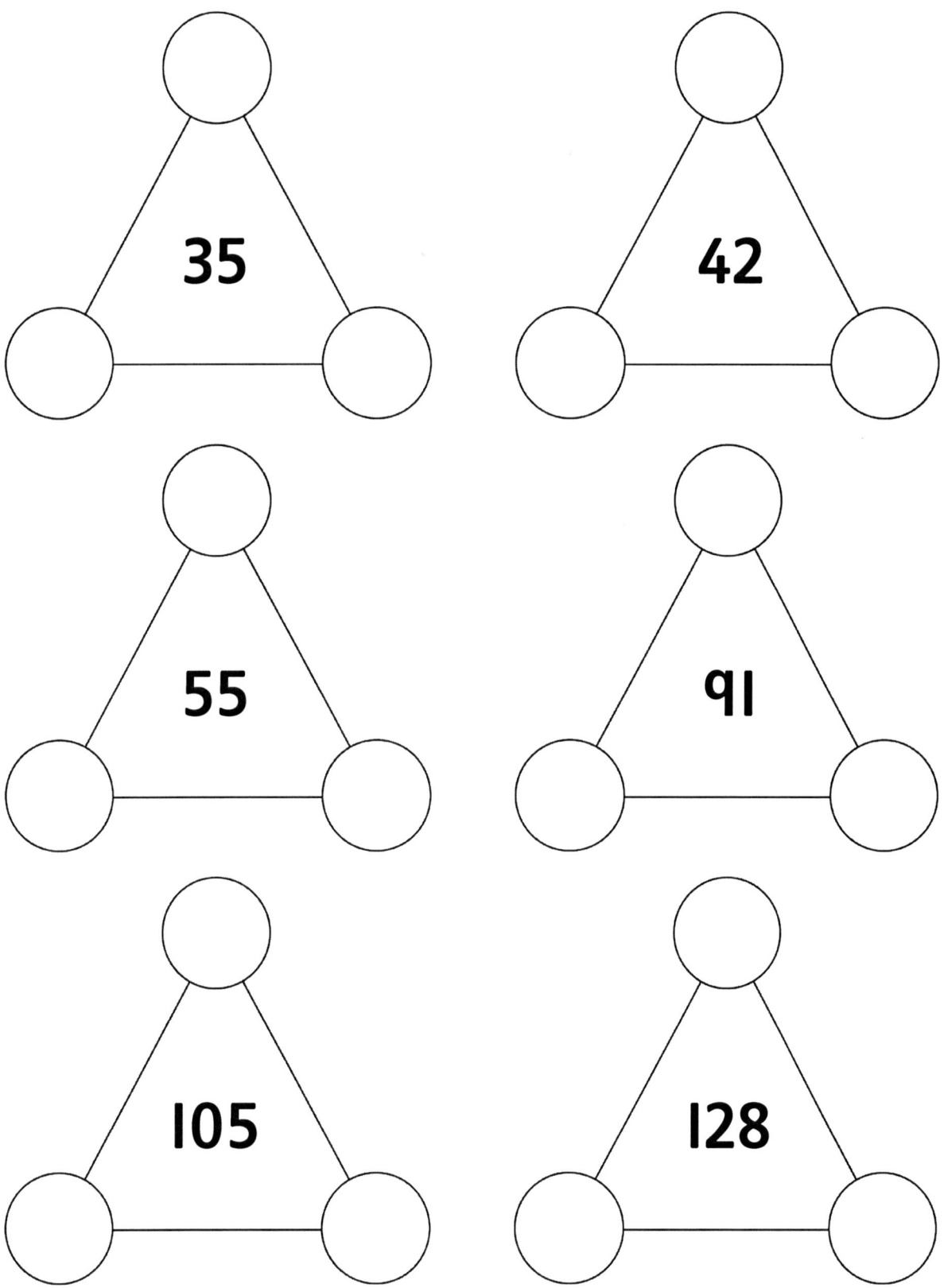

Buying teddies

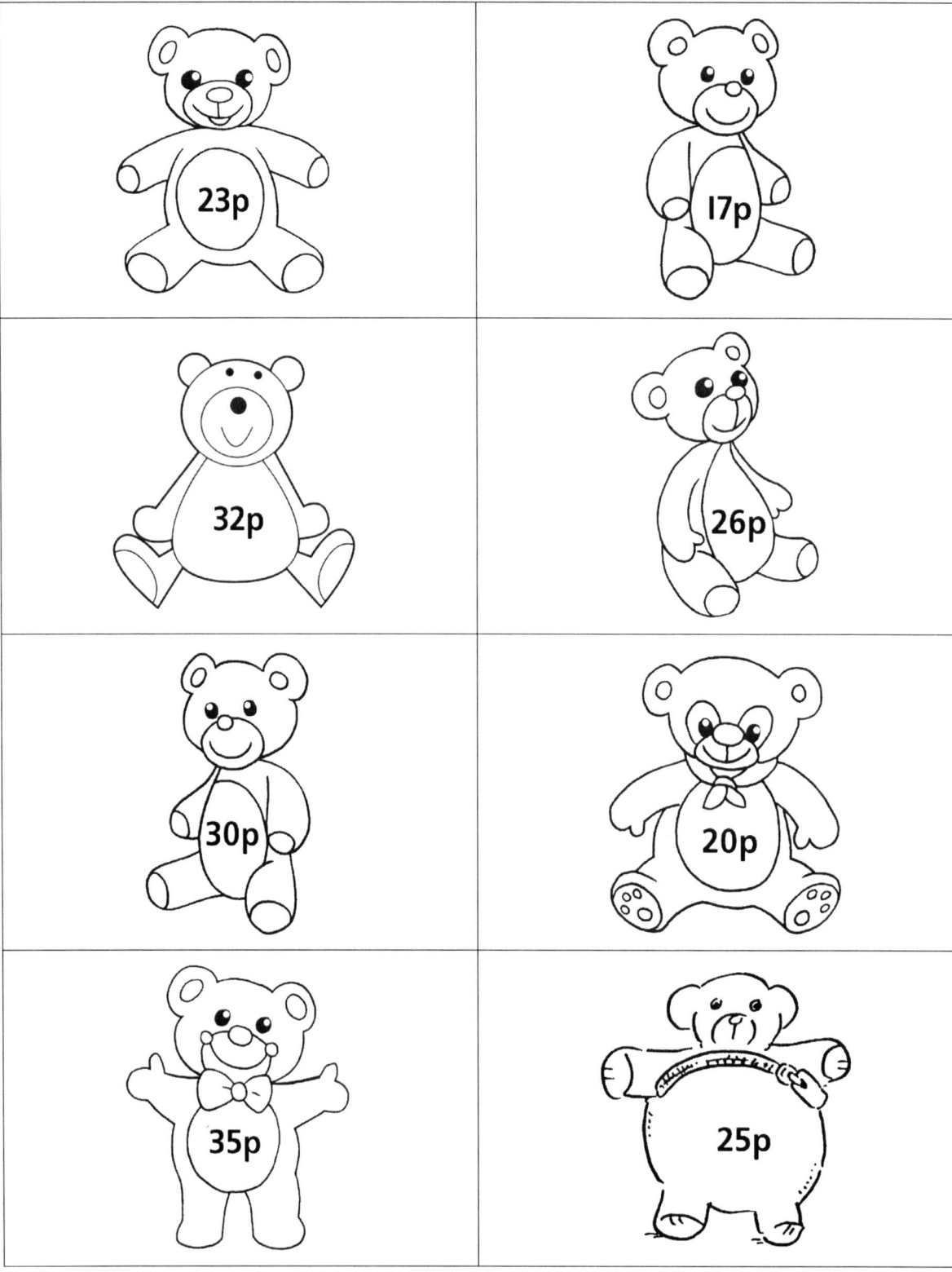

Money problems

A lolly costs 2p.

1. How much would it cost to buy:
    a) 6 lollies
    b) 9 lollies
    c) 13 lollies?

2. How many lollies can you buy with:
    a) 16p
    b) 24p
    c) 27p?

A pencil costs 5p.

3. How much would it cost to buy:
    a) 7 pencils
    b) 11 pencils
    c) 15 pencils?

4. How many pencils can you buy with:
    a) 45p
    b) 65p
    c) 32p?

A rubber costs 10p.

5. How much would it cost to buy:
    a) 6 rubbers
    b) 12 rubbers
    c) 21 rubbers?

6. How many rubbers can you buy with:
    a) 70p
    b) 90p
    c) 72p?

Addition questions

Answer these questions.

1. 37 + 54 = ⬚

2. 68 + 43 = ⬚

3. 132 + 64 = ⬚

4. 148 + 36 = ⬚

5. 176 + 47 = ⬚

6. 153 + 162 = ⬚

Number problems

Solve the following problems and record your solutions.

1. Peter has 8 marbles, George has 6 marbles. The marbles are green. How many marbles altogether?

2. Rose has 20p. She buys a blue pencil for 8p. How much does she have left?

3. Ahmed has 5 sweets, Gill has 9 sweets, David has 4 sweets. How many sweets do they have altogether?

4. Hassan has 15 marbles. He gives 8 to his friend. How many does he have left?

5. Merinda's pencil is 14 cm long, Jacob's pencil is 6 cm long. How much longer is Merinda's pencil than Jacob's?

---

1. Peter has 68 marbles, George has 27 marbles. The marbles are green. How many marbles altogether?

2. Rose has £1.35p. She buys a red pen for 29p. How much money does she have left?

3. Ahmed has 64 sweets, Gill has 49 sweets, David has 34 sweets. How many sweets do they have altogether?

4. Hassan has 76 marbles. He gives 28 to his friend. How many does he have left?

5. Merinda's skipping rope is 93 cm long, Jacob's skipping rope is 66 cm long. How much longer is Merinda's skipping rope than Jacob's?

Properties of shapes

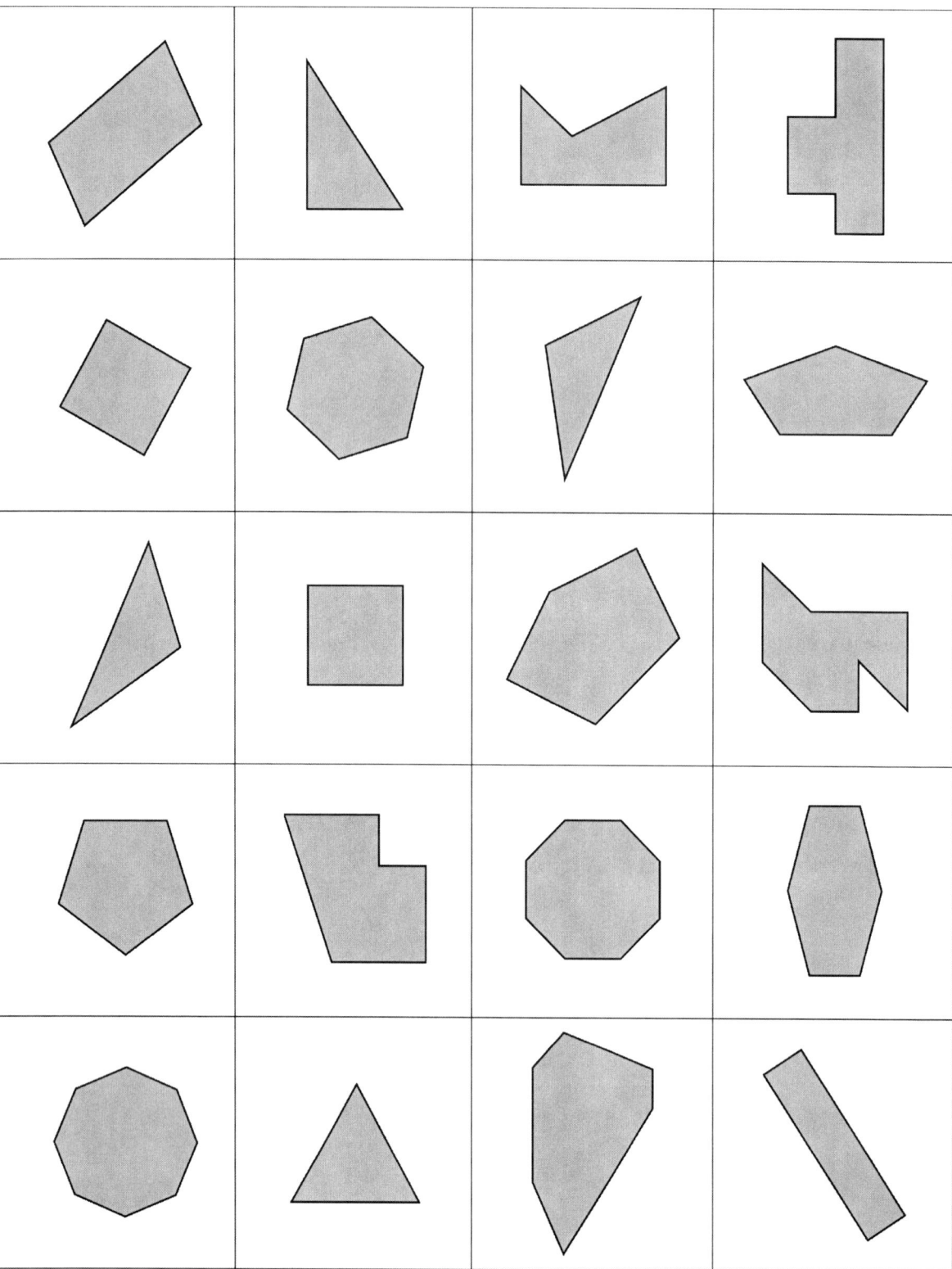

Positions

Routes

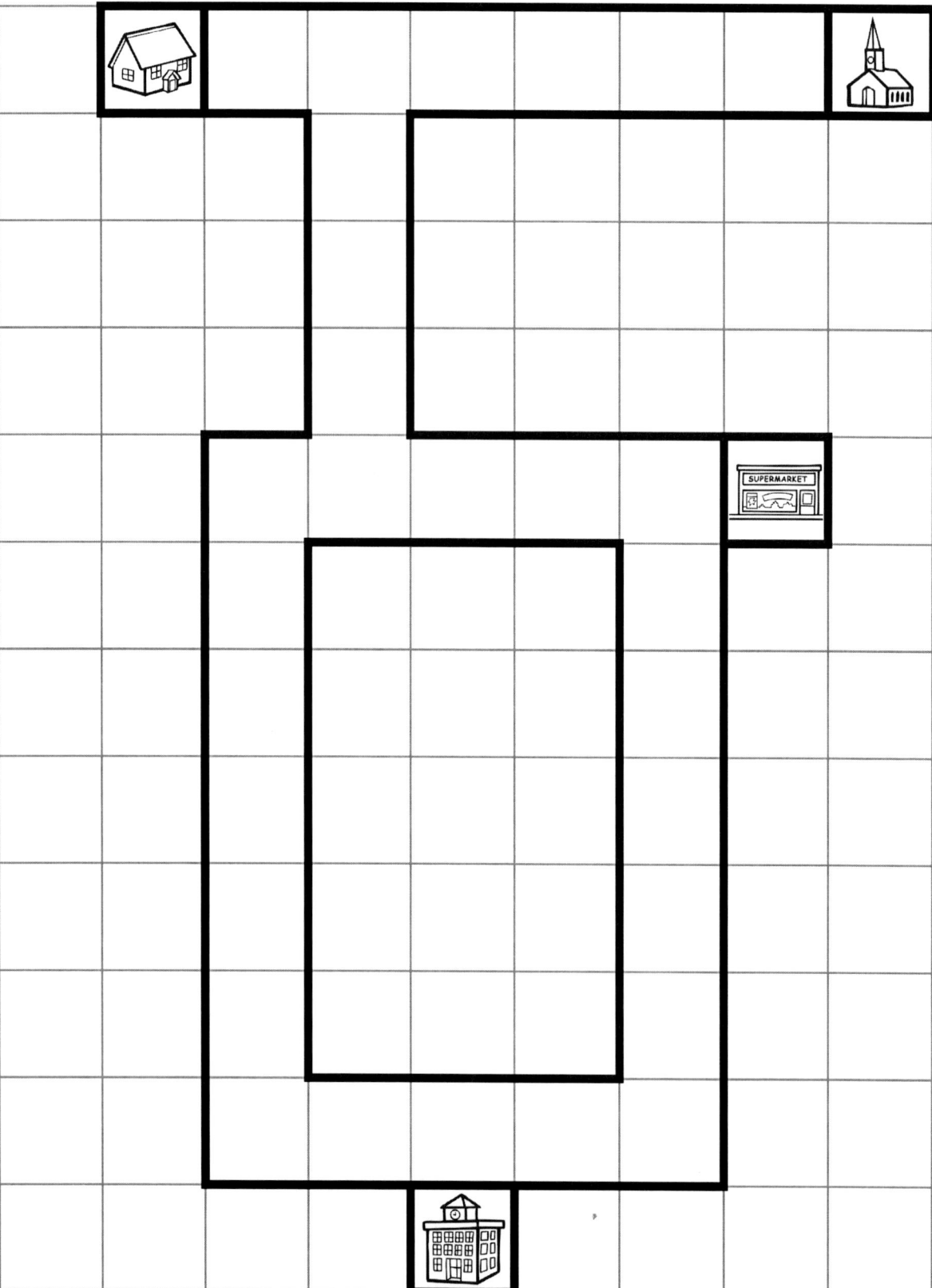

Measuring length

1. Measure these lines to nearest centimetre/half centimetre:

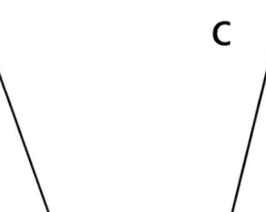

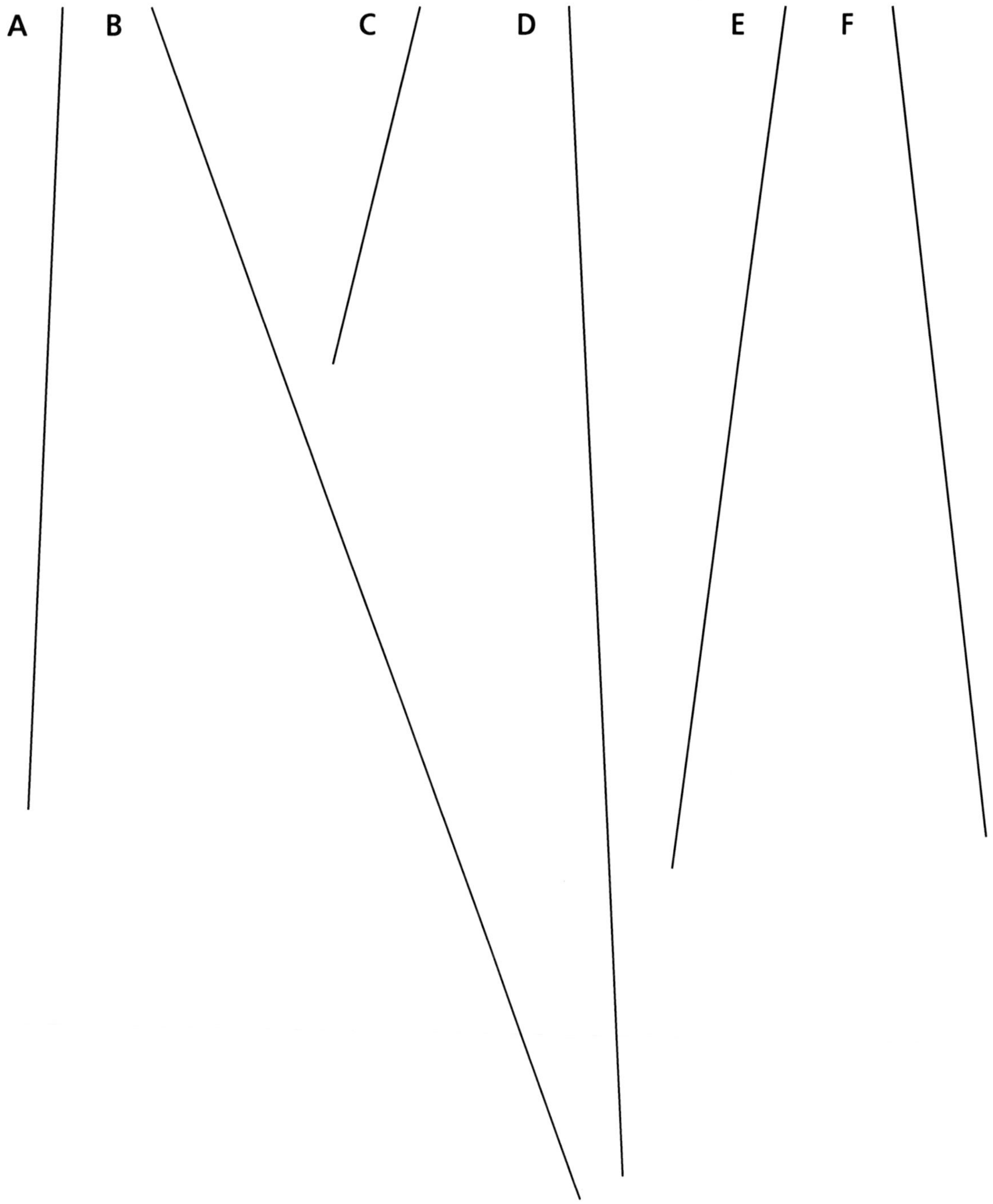

2. On a plain piece of paper draw lines of these lengths:

A 11 cm

B $7\frac{1}{2}$ cm

C $13\frac{1}{2}$ cm

D 21 cm

Finding the right angles in shapes

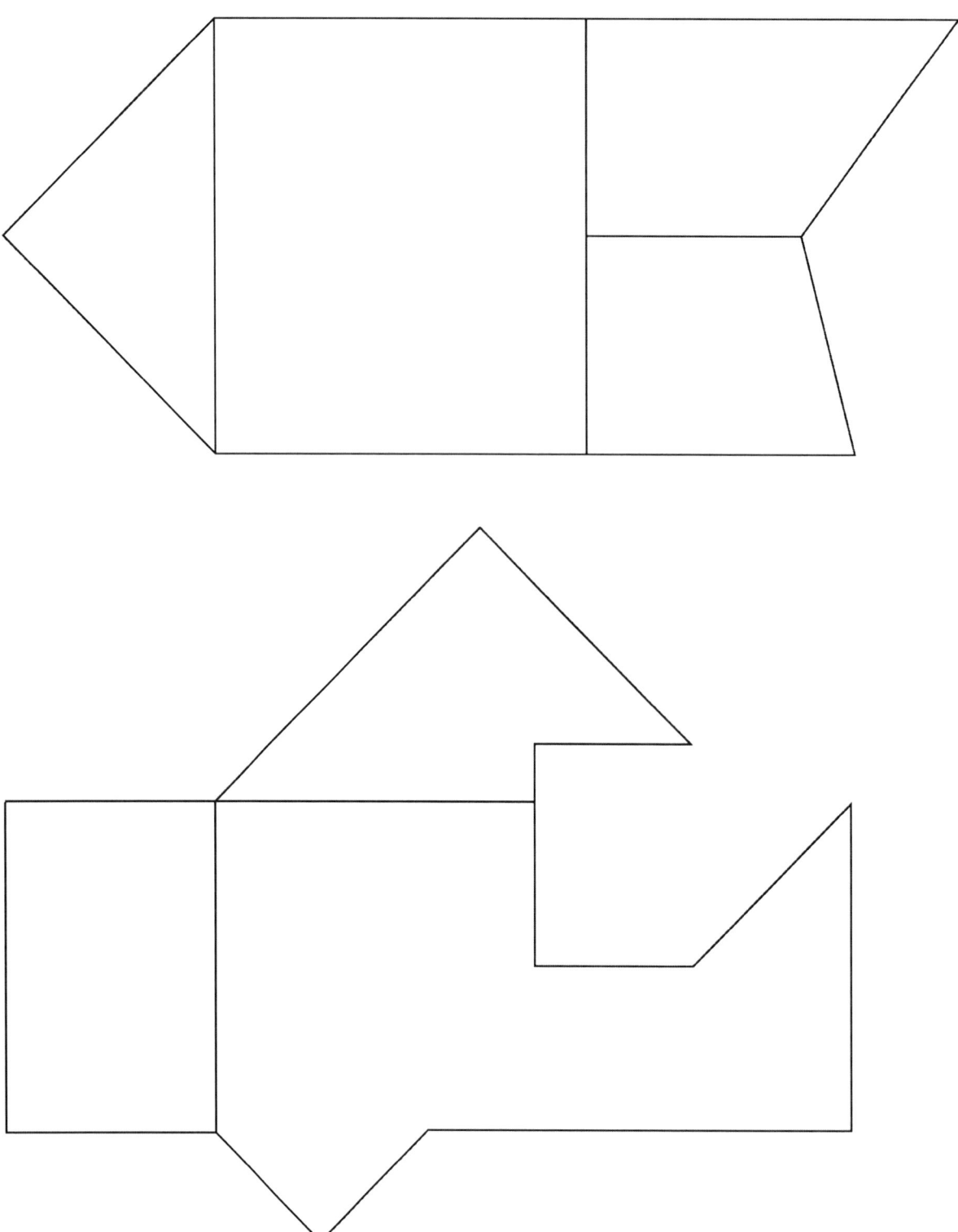

Reading scales

Collecting data

Ann	Football
Beth	Tennis
Cara	Cricket
Dee	Tennis
Erin	Rounders
Farida	Basketball
Greg	Cricket
Hassan	Football
Hilda	Rounders
Inaz	Football
Jamie	Basketball
Joy	Rounders
Karla	Football
Kelly	Cricket
Layla	Football
Luz	Rounders
Mark	Football
Mikela	Basketball
Nathan	Rounders
Omar	Football
Penny	Cricket
Quentin	Rounders
Raz	Tennis
Tamsin	Cricket
Una	Football
Valerie	Cricket
Wes	Tennis
Will	Football

Tally chart

Basketball	
Cricket	
Football	
Rounders	
Tennis	

Processing and representing data

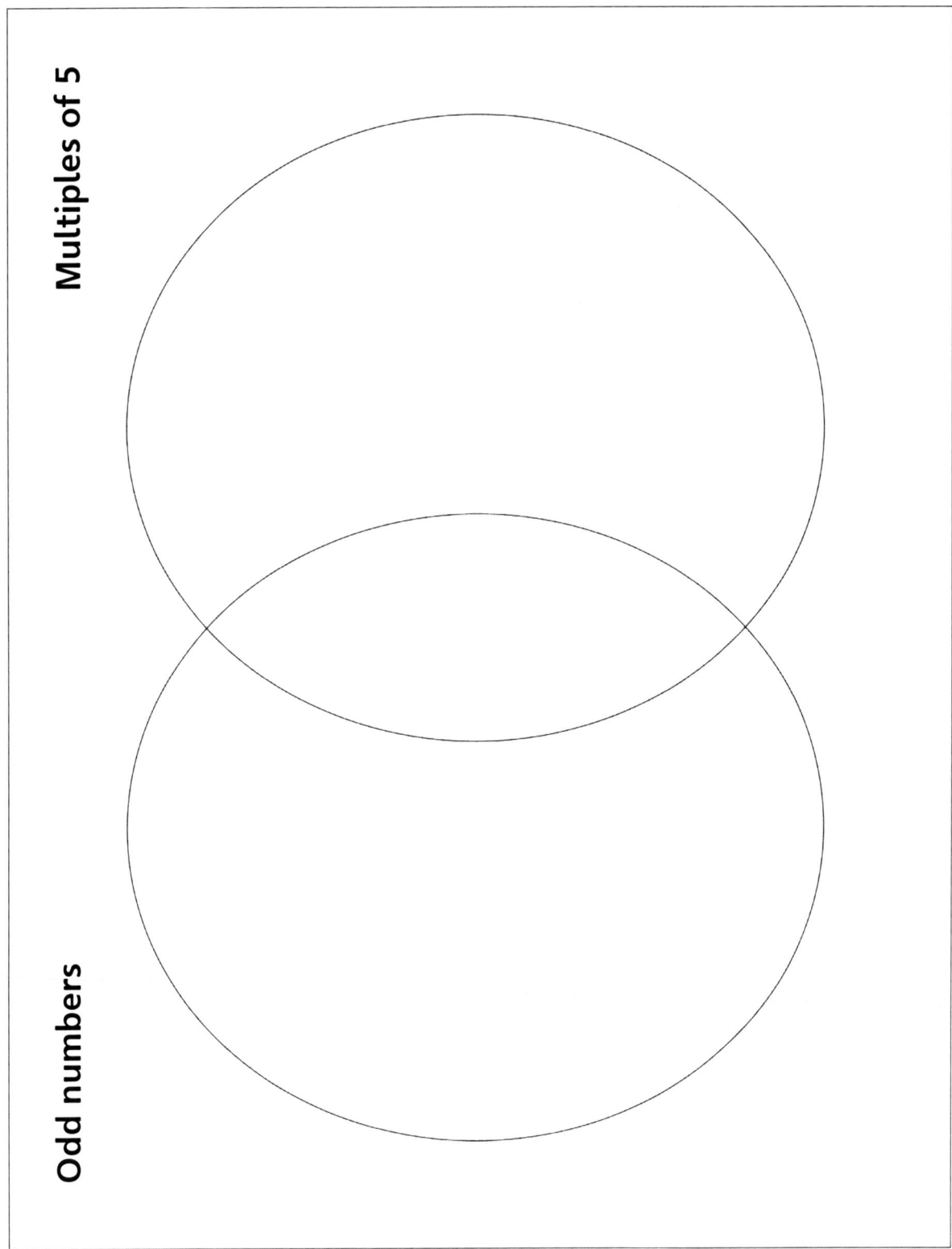

Multiples of 5

Odd numbers

Block graph (1)

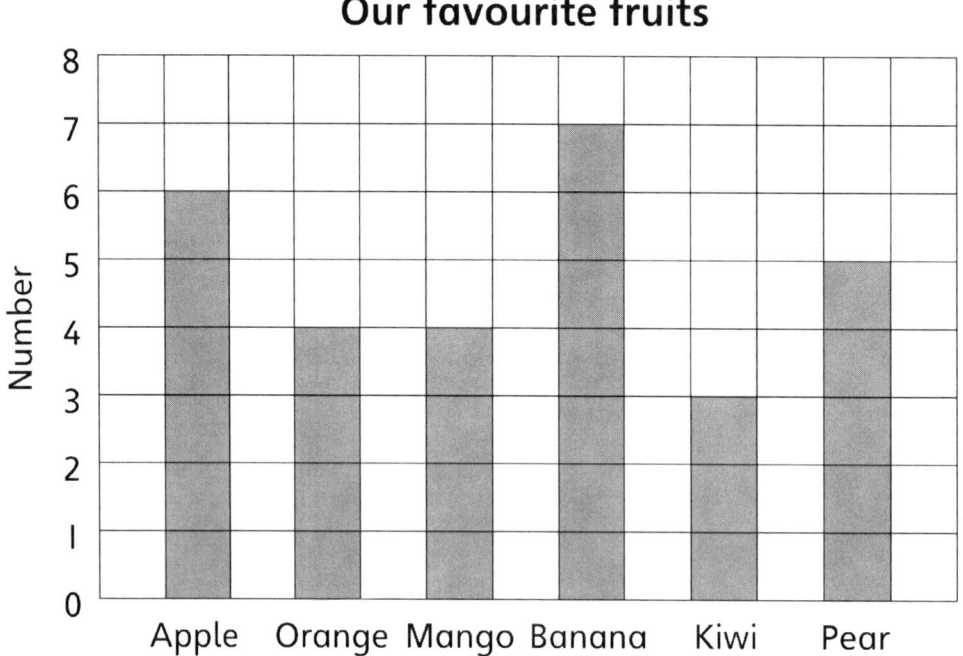

Our favourite fruits

1. What is the most popular fruit? _____

2. What is the least popular fruit? _____

3. How many children chose Kiwi and Pear altogether?   ____

4. How many children chose Apple and Banana
   altogether?   ____

5. How many more children chose Banana than Mango? ____

6. How many fewer children chose Kiwi than Apple?   ____

7. How many children in the class altogether?   ____

8. Put the fruits in order from the least to most popular.

_____

_____

_____

Block graph (2)

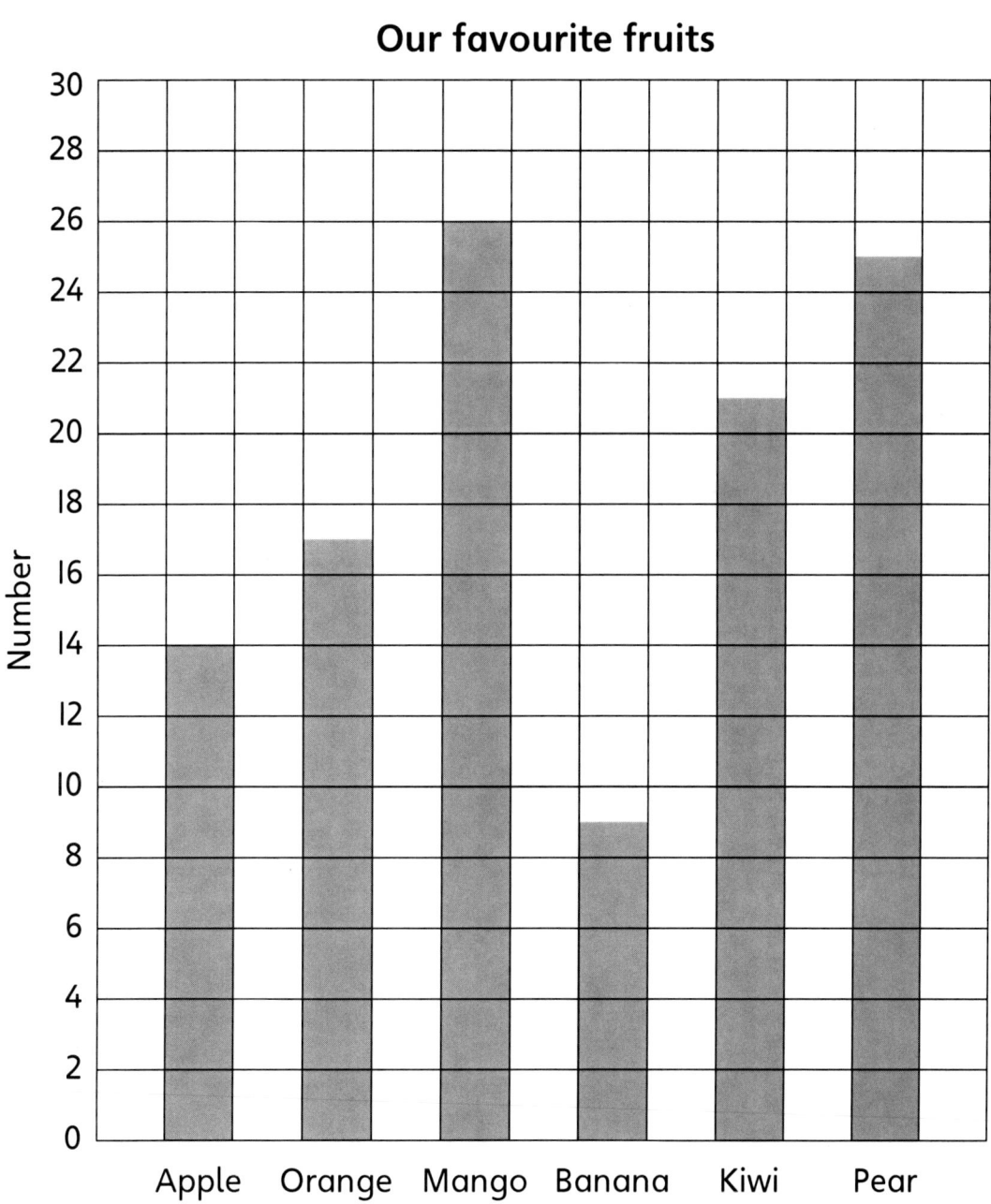

Our favourite fruits

Write your own questions about this graph. Use the questions on Pupil Sheet 30.1 to help you.

Pearson Education Limited is a company incorporated in England and Wales, having its registered office at Edinburgh Gate, Harlow, Essex, CM20 2JE. Registered company number: 872828

www.pearsonschools.co.uk

Text © Pearson Education 2010

First published 2010

14 13 12 11 10

10 9 8 7 6 5 4 3 2 1

British Library Cataloguing in Publication Data
A catalogue record for this book is available from the British Library.

ISBN 978 0 435041 41 0

Editorial team: Jenny Penfold, Rhian McKay, Katie Frederick.

Designed by Mike Brain Graphic Design Limited.

Original illustrations © Pearson Education, 2010.
Illustrated by Andy Robb (Beehive Illustration).

Cover design by Clive Goodyer.

Cover illustration © Clive Goodyer.

Printed in the UK by Henry Ling.

Acknowledgements
Every effort has been made to contact copyright holders of material reproduced in this book. Any omissions will be rectified in subsequent printings if notice is given to the publishers.

Some material has been adapted from *Abacus Evolve Assessment Kit Year 2*, previously published by Pearson Education.